Forest and Plant Bioproducts Division

Presentations at the 2008 AIChE Annual Meeting

Philadelphia, Pennsylvania
16 - 21 November 2008

ISBN: 978-1-61567-230-1

Printed from e-media with permission by:

Curran Associates, Inc.
57 Morehouse Lane
Red Hook, NY 12571

Some format issues inherent in the e-media version may also appear in this print version.

Printed by Curran Associates, Inc. (2009)

For permission requests, please contact AIChE
at the address below.

AIChE
3 Park Avenue
New York, NY 10016-5991

Phone: (203) 702-7660
Fax: (203) 775-5177

www.aiche.org

Additional copies of this publication are available from:

Curran Associates, Inc.
57 Morehouse Lane
Red Hook, NY 12571 USA
Phone: 845-758-0400
Fax: 845-758-2634
Email: curran@proceedings.com
Web: www.proceedings.com

TABLE OF CONTENTS

EFFECT OF CLAY ON EXTRUSION FOAMING OF WF/PP/CLAY COMPOSITES USING N_2

Y.H. Lee[1], T. Kuboki[2], C.B. Park[2] and M. Sain[1]
1. Centre for Biocomposites and Biomaterials Processing
Faculty of Forestry, University of Toronto, Toronto, ON, Canada
2. Microcellular Plastics Manufacturing Laboratory
Department of Mechanical and Industrial Engineering, University of Toronto
Toronto, ON, Canada

Abstract

The paper investigated the effect of clay on extrusion foaming of polypropylene (PP)/clay composites and wood fiber (WF)/PP/clay composites that contained a small amount of clay, up to 5 wt%. The composites were prepared by a melt blending masterbatch process. The clay dispersion in the composites was characterized by a X-ray diffractometer while the viscosity of the composites was evaluated by a rotational rheometer. The foaming behaviour of the composites was examined using an extrusion foaming system, in which N_2 was used for foams as a physical blowing agent. The results suggested that the organoclay provided higher cell density for PP/clay composites than the natural clay, and the addition of organoclay was more effective in improving the cell morphology (decrease of cell size and more uniform cell size) for the WF/PP/clay composites with lower fiber loading.

Introduction

Wood fiber plastic composites (WPC) have received considerable attention lately within the automotive, construction, and furniture industries due to their low cost and recyclability. However, the density of WPC is higher than those of plastics and wood, which is one of the major obstacles that prevent the expansion of their application.

Foaming technology can decrease the density of WPC, but it is not an easy task to maintain or increase their functional properties when the density is reduced. Since the properties of foams can be improved with decreasing cell size and increasing uniformity [1,2], it is essential to develop WPC foams with reduced cell size and narrower cell-size distribution.

An addition of layered silicate particles (i.e., clay) may be one of the ways to decrease cell size and achieve more uniform cells in WPC. It has been reported that the addition of clay improved cell morphology of various polymers, such as low-density polyethylene (LDPE) [3], high-density polyethylene (HDPE) [4], polystyrene (PS) [5,6], polylactide (PLA) [7], polyamide [8], polypropylene (PP) [9], and maleated PP [10]. However, very few literatures are available on studying the effect of clay on the foaming behavior of WPC. Guo et al. [11] reported that the addition of clay reduces the cell size and increases the cell density of wood fiber/metallocene polyethylene composites in extrusion foaming using a chemical blowing agent.

In this paper, two types of clay (natural clay and organoclay) were blended with PP, and the foaming behaviors of PP/clay composites were examined in extrusion using a physical blowing agent, N_2. Since the organoclay was found to be more effective in improving cell morphology of PP (see the

"Results and Discussion" section), it was employed to investigate foaming behavior of WF/PP/clay composites.

Experimental

Materials

PP (PP1024E4, ExxonMobile Chemicals), with a density of 0.91 g/cm^3 and a melt flow index of 12.5 g/10min, was used as a polymer matrix. Wood fiber (2020, American Wood Fibers) of which size is in the range of 250-850 μm was used as a reinforcement. Two types of clay were employed: natural clay (Cloisite Na+, Southern Clay Products) and clay modified with dimethyl dihydrogenated tallow alkyl ammonium (Cloisite 20A, Southern Clay Products). Maleated PP (PP-MA, Polybond 3150, 1.0-1.5 wt% of MAH, MI 60g/10min, Chemtura) was used to improve the adhesion between the hydrophobic PP and the hydrophilic wood fiber or layered silicate. The wood fiber was dried at 70°C for 24 hours in a convection oven prior to mixing in order to remove moisture in the wood fiber.

Preparation of Composites

A masterbatch was prepared by melt-blending PP, PP-MA, and clay (either natural clay or organoclay) with a mixing ratio of 8:1:1 using a high-intensity kinetic mixer (W&P Gelimat). The mixer was run at 3200 rpm with a tip speed of 23 m/s. This mixing encourages a chemical interaction between the hydroxyl groups present on the nanoclay and the PP-MA to disperse the clay particles uniformly throughout the polymer matrix. In addition, the higher melt viscosity of PP offers higher shear stress on the clay. The melt blends were discharged at a preset temperature of 185°C, and then cooled down to room temperature. The solidified blends were granulated using a granulator (S-10-9, C.W. Brabender). The masterbatch granules were dry-blended with PP and wood fiber to yield specific material compositions. Again, the blends were melt-blended and granulated using the same equipment.

The wood fiber and organoclay contents varied from a range of 0 to 50 wt% and 0 to 5 wt%, respectively, but the PP-MA content was fixed at 5 wt%. The formulation of the composites is summarized in Table 1.

X-ray Diffraction

X-ray Diffraction (XRD) was conducted with a Siemens D5000 diffractometer that uses Cu Kα radiation (1.548Å) and a Kevex solid-state detector. Measurements were performed at 50 kV and 35 mA. The data was recorded in the reflection mode using a STEP scan method; the step size was 0.02 degrees and the counting time was 2.0 seconds per step.

Viscosity

The samples were subjected to oscillatory shear in a dynamic stress rheometer (Rheometric Scientific SR-200) with parallel plates (25 mm in diameter) and at a gap of 1.0 mm. In all cases, the dynamic stress sweeps were initially performed to define the limits of the linear viscoelastic regime. The dynamic frequency sweep experiments were then conducted to measure the complex viscosity (η^*) over a frequency range of 0.1-100 rad/s at 200°C. All measurements were performed under N_2 to prevent polymer degradation or moisture absorption.

Extrusion Foaming

The PP/clay or WF/PP/clay pellets were processed in a single-screw, three-zone extruder with a gas injection system. In this study, nitrogen (N_2) was used as an environmentally friendly physical blowing agent. The foamability of the composites was studied in the extrusion using N_2, at 0.3 wt%. The temperature in this extruder maintained at 180°C. A static mixer, a heat exchanger, and a filament die (L/D: 1.5"/0.05") were attached to the downstream of the barrel. Their temperatures were set at 170°C. The die temperature was reduced gradually from 170°C to 135°C in order to prevent gas loss in the extrudate which would consequently reduce the composite density. The foam samples were collected at each stabilized processing condition and characterized for foam density and cell morphology, such as cell size, cell size distribution, and cell density. Each sample was dipped in liquid nitrogen and fractured. The fractured surface was then gold-coated using a sputter coater (E 50000C PS3). The cellular morphology was examined using scanning electron microscopy (SEM, JEOL JSM-6060).

Results and Discussion

PP/Clay Composites

Clay Dispersion and Viscosity

Figure 1 (a) shows the X-ray diffraction patterns of PP/clay composites with natural clay. The interlayer spacing of the pure natural clay before compounding is 1.1 nm. After compounding, the X-ray diffraction peaks of the composites with 3 and 5 wt% clay as well as the master batch shifted to higher angles. The results suggest that the interlayer spacing slightly decreased to around 1.0 nm for all the composites. Figure 1 (b) shows the X-ray diffraction patterns of PP/clay composites with organoclay. The interlayer spacing of pure organoclay before compounding is 2.6 nm. After compounding, the X-ray diffraction peaks of all the composites were broadened, which suggests that the interlayer thickness was disordered.

Figure 2 shows the complex viscosity of PP/clay composites at different frequencies. This figure indicates that the addition of natural clay created few changes to the viscosity of PP. However, the addition of organoclay increased viscosity of PP at a low frequency level, which suggests that there is an interaction between the organoclay and PP.

Foaming Behavior

Figure 3 shows the foam density of PP/clay composites at different die temperatures. The density of PP/clay composites was decreased to around 0.6 g/cm^3 by foaming. Also, it was observed that the foam density of the composites was insensitive to the die temperature, while the clay type and content little affected the foam density. However, as shown in Figure 4, the cell density increased with the increase of clay content for both natural clay and organoclay composites. Furthermore, the organoclay composites showed higher cell density than natural clay composites.

It is speculated that the addition of clay particles increased the nucleation rate of bubbles in keeping with the principles of heterogeneous nucleation [12]. In heterogeneous nucleation, the activation energy barrier to nucleation is reduced in the presence of filler particles. As a result, the addition of clay particles increases cell nucleation. The higher cell density in the organoclay composites than natural clay composites can be explained by the larger interfacial area between clay and PP matrix. The organoclay composites had clay with intercalated structure and larger number of particles, which provided more sites for cell nucleation. In addition to the enhanced cell nucleation, it is surmised that the larger interfacial area suppressed the cell growth. A higher fraction of polymer chains is confined by clay, and thus their mobility is restricted, which increased melt viscosity of the composites (see Figure 2).

WF/PP/Clay Composites

The results from PP/clay composites indicated that organoclay improve cell morphology of PP more than natural clay. Therefore, organoclay was selected as the clay for WF/PP/clay composites.

Clay Dispersion and Viscosity

Figure 5 shows the X-ray diffraction patterns of WF/PP/clay composites as well as pure organoclay. After compounding, X-ray diffraction peaks of all the composites shifted to lower angles (3.6 nm or so). The result indicates that the interlayer spacing was increased by the intercalation of polymer, but the organoclay still retains an ordered structure. It is expected that the compounding of viscous wood fibre composites facilitated the intercalation of the organoclay.

Figure 6 shows the complex viscosity of WF/PP/clay composites at different frequencies. For both composites with 25 wt% and 50 wt% wood fiber composites, the viscosity of the composites increased with the increase of clay content. It is also noted that the viscosity of the composites increased with the wood fiber content (see Figure 2 for 0 wt% wood fiber composites).

Foaming Behavior

Figure 7 shows the foam density of WF/PP/clay composites with 25 wt% wood fiber at different die temperatures. Similar to PP/clay composites, foam density of the composites was insensitive to the die temperature and clay content little affected the foam density.

Figure 8(a) shows the cell size and cell size distribution of WF/PP/clay composites with 25 wt% wood fiber at a die temperature of 140°C. With the increase of clay content, the average cell size of the composite decreased, and the cell size became more uniform, which is indicated by the steeper curves on the graph. Analogous to the PP/organoclay composites, the improvement in cell morphology of WF/PP/clay composite foams with the increase of organoclay can be explained by two effects: (i) clay particles act as nucleating agents, thus promoting heterogeneous nucleation, and (ii) the melt viscosity of the matrix increases with an increase in clay content (see Figure 6), generating higher resistance to cell growth in the foaming processes.

However, for WF/PP/clay composite with 50 wt% wood fiber (Figure 8(b)), there was no significant improvement in cell size and cell size uniformity with the addition of organoclay (up to 5 wt%). This can be attributed to the higher interfacial area between wood fiber and PP matrix for cell nucleation as the wood fiber content is increased (from 25 wt% to 50 wt%), which decreased the effectiveness of the organoclay particles as nucleating agents.

Conclusions

The foaming behavior of PP/clay composites and WF/PP/clay composites were examined in extrusion. The results from the PP/clay composites suggest that the cell density increased with the increase of clay content for both natural clay and organoclay composites. However, composites with organoclay showed higher cell density than those with natural clay. The results from WF/PP/clay composites suggest that, in the composites with 25 wt% wood fiber, the average cell size decreased and the cell size became more uniform with the increase of organoclay content while cell size and cell size uniformity of the composites with 50 wt% wood fiber increased slightly with the addition of organoclay.

Acknowledgement

This work was supported financially by AUTO21, the Consortium for Cellular and Microcellular Plastics (CCMCP), and the Natural Sciences and Engineering Research Council of Canada (NSERC). We also wish to express our thanks to ExxonMobile Chemical Company, American Wood Fibers and Chemtura for supplying materials.

References

1. D.I. Collias, D.G. Baird, and R.J.M. Borggreve, *Polymer*, **35**, 3978 (1994).
2. D.I. Collias, D.G. Baird, *Polym. Eng. Sci.*, **35**, 1167 (1995).
3. Y.H. Lee, K.H. Wang, C.B. Park, and M. Sain, *J. Appl. Polym. Sci.*, **103**, 2129 (2007).
4. Y.H. Lee, C.B. Park, K.H. Wang, and M.H. Lee, *J. Cell. Plast.*, **41**, 487 (2005).
5. X. Han, C. Zeng, L.J Lee, D.L. Tomasko, and K.W. Koelling, *PPS-18*, PO23 (2002).
6. X. Han, C. Zeng, L.J. Lee, D.L. Tomasko, and K.W. Koelling, *SPE ANTEC Tech. Papers,* 1915 (2002).
7. Y. Fujimoto, S.S. Ray, M. Okamoto, A. Ogami, K. Yamada, and K. Ueda, *Macromol. Rapid. Commun.*, **24**, 457 (2003).
8. H. Kharbas, P. Nelson, M. Yuan, S. Gong, and L.S. Turng, *Polym. Comp.*, **24**, 655 (2003).
9. W.G. Zheng, Y.H. Lee, and C.B. Park, *J. Appl. Polym. Sci., submitted, September* (2007).
10. P.H. Nam, P. Maiti, M. Okamoto, T. Kotaka, T. Nakayama, M. Takada, M. Ohshima, A. Usuki, N. Hasegawa, and H. Okamoto, *Polym. Eng. Sci.*, **42**, 1907 (2002).
11. G. Guo, K.H. Wang, C.B. Park, Y.S. Kim, and G. Li, *J. Appl. Polym. Sci.*, **104**, 1058 (2007).
12. J.S. Colton, and N.P. Suh, *Polym. Eng. Sci.*, **27**, 500 (1987).

Table 1. Material Information of the Composite Materials Prepared in This Study.

No	Wood Fibre (wt%)	PP (wt%)	PP-MA (wt%)	Clay (wt%)	
				Organoclay	Natural clay
1	0	95	5	0	0
2	0	92	5	3	0
3	0	90	5	5	0
4	0	92	5	0	3
5	0	90	5	0	5
6	25	70	5	0	0
7	25	67	5	3	0
8	25	65	5	5	0
9	50	45	5	0	0
10	50	42	5	3	0
11	50	40	5	5	0

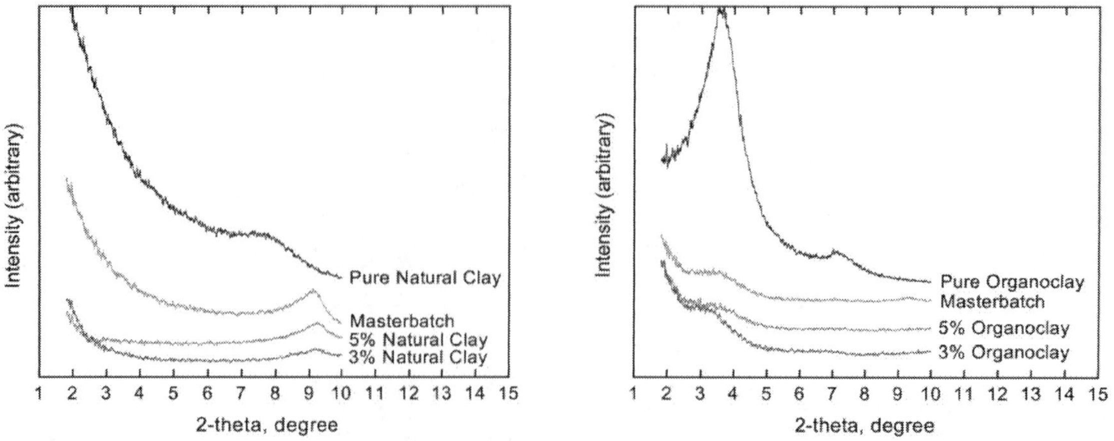

Figure 1. XRD Patterns of PP/Clay Composites: (a) Natural clay and (b) Organoclay

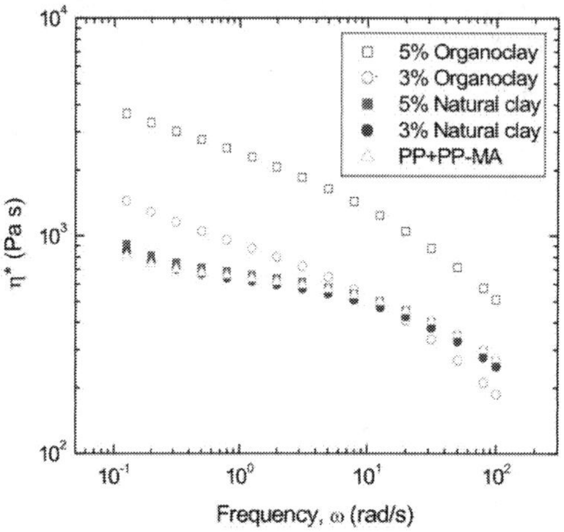

Figure 2. Complex Viscosity (η*) of PP/Clay Composites as a Function of Frequency

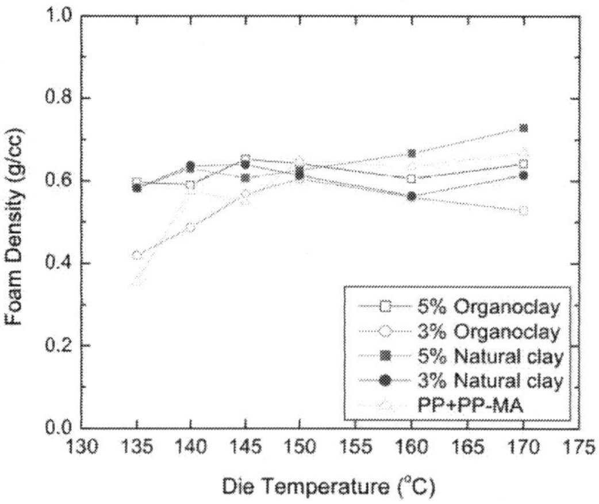

Figure 3. Foam Density of PP/Clay Composites as a Function of Die Temperature

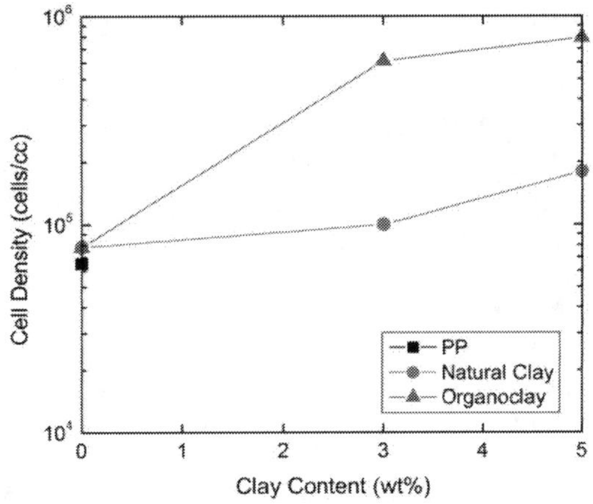

Figure 4. Cell Density of PP/Clay Composites as a Function of Clay Content

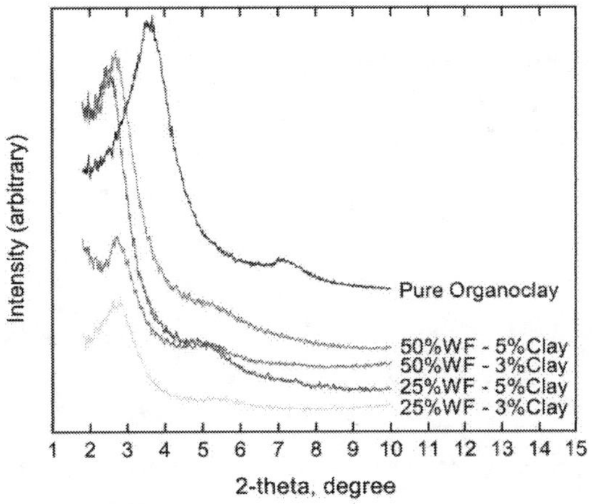

Figure 5. XRD Patterns of WF/PP/Clay Composites

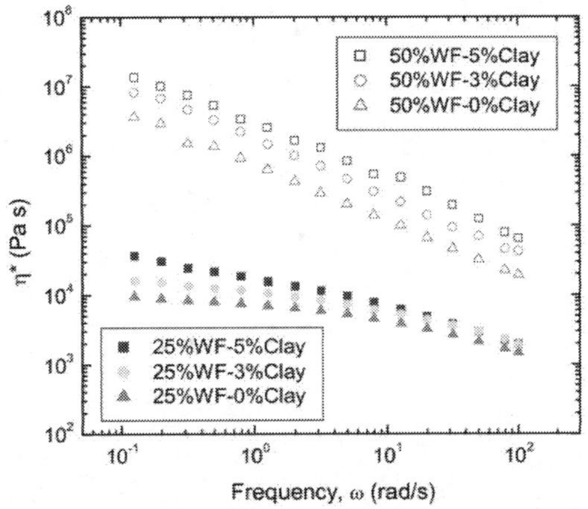

Figure 6. Complex Viscosity (η*) of WF/PP/Clay Composites as a Function of Frequency

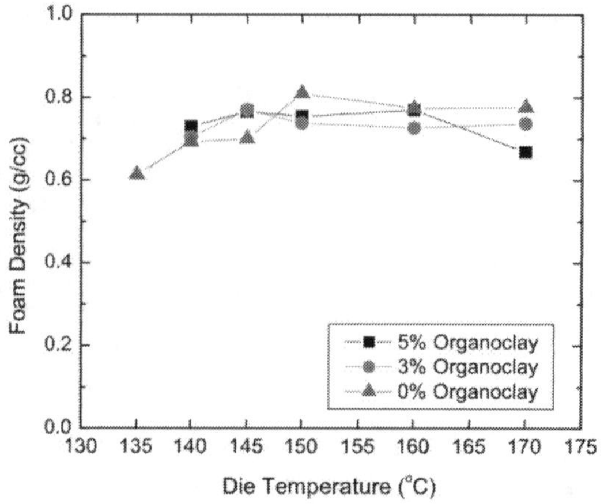

Figure 7. Foam Density of WF/PP/Clay Composites with 25 wt% Wood Fiber as a Function of Die Temperature

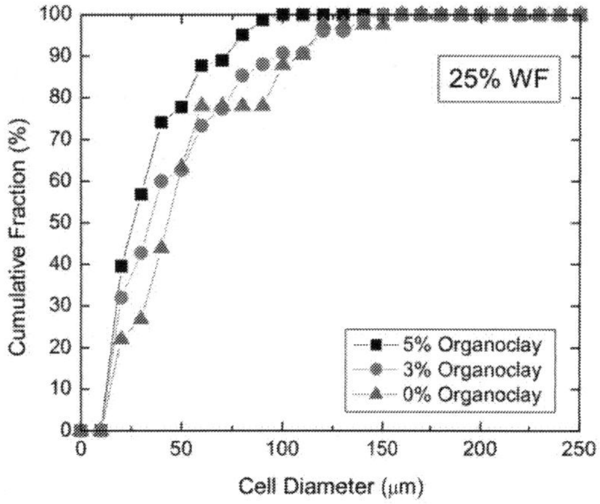

(a)

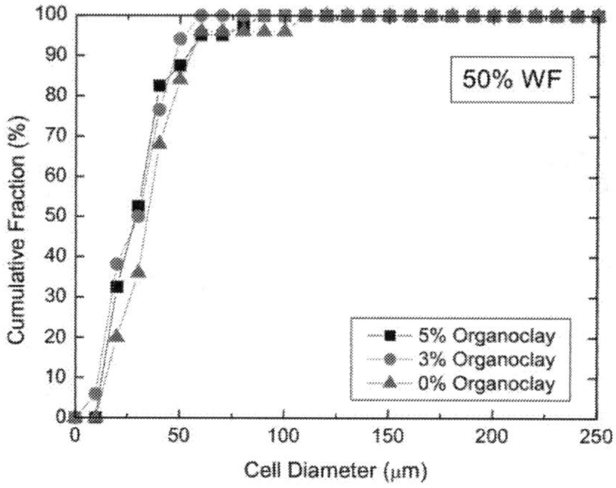

(b)

Figure 8. Cumulative Cell Size Distribution of WF/PP/Clay Composite Foams at a Die Temperature of 140°C: (a) 25 wt% Wood Fiber and (b) 50 wt% Wood Fiber

395b Fundamental Design of Bio-Based Polymers: Solutions to Global Warming

Richard P. Wool, Chemical Engineering, University of Delaware, 150 Academy Street, Colburn Lab, Newark, DE 19716

As a viable solution to global warming, bio-based materials are being designed to support the green renewable energy infrastructure with wind turbines, solar integrated energy efficient housing, hydro turbines and hydrogen storage. Recent advances in genetic engineering, composite science, and natural fiber development offer significant opportunities for new, improved green materials from renewable resources that are optionally recyclable, biocompatible and biodegradable, thereby enhancing global sustainability. When such biobased resins are combined with natural fibers (plant and poultry) starch and lignin, new low-cost composites, pressure sensitive adhesives, elastomers and foams are produced that are economical in many high-volume applications. The Twinkling Fractal Theory of the Glass Transition and Yield and the Percolation Theory of Fracture and polymer melt rheology are used to optimize the molecular design and properties of new bio-based materials. These high performance composites can be used in energy efficient solar integrated roofs, wind foil blades, hurricane resistant housing, sub-aqua hydro turbines and hydrogen storage, in addition to agricultural equipment, automotive sheet molding compounds, civil and rail infrastructures, marine applications, electronic materials, and sports equipment. The development of biobased materials is consistent with the principles of Green Chemistry and Engineering, which pertain to the design, commercialization and use of processes and products that are technically and economically feasible while minimizing the generation of pollution at the source and the risk to human health and the environment.

395c New Biobased Materials from Co-Products

Amar K. Mohanty, Department of Plant Agriculture & School of Engineering, Crop Science Building, University of Guelph, Guelph, ON N1G 2W1, Canada

The generation of biofuels, biochemicals and biomaterials from renewable biomasses are getting a great thrust as potential substitutes and supplements to the petroleum based counterparts. The biofuels like bioethanol, biobutanol and biodiesel derived from plant and forest resources are getting accelerated momentum across the world. All the components associated with the alternative fuels should be sustainable for a well-balanced bioeconomy. The downstream coproducts and byproducts from these emerging alternative fuel industries need value-added uses. In lingo-cellulosic biomass conversion into ethanol, the substantial quantity of low or negative value leftovers such as lignin, carbon dioxide and residues are finding great attention in their value-added uses. In ligno-cellulosic ethanol industries; ethanol and lignin are produced at 1:1 on a quantum basis in general. Similarly in corn ethanol dry-milling process, one bushel of corn produces 1/3rd as ethanol, 1/3rd as carbon dioxide and rest 1/3rd as Distillers' Dried Grains as Solubles (DDGS). Soy meal (SM), canola meal (CM), jatropha meal (JM) and crude glycerol are coproducts and byproducts of the biodiesel industry. This presentation will highlight the research results, current status, opportunities, and challenges on engineering uses of the downstream products as plastic resins, biodegradable plastics, bioadhesives, biorubber and biocomposites.

395d Effects of Adding Chain Extender on the Properties of Injection-Molded Solid and Microcellular Polylactide

Shaoqin (Sarah) Gong[1], Lih-Sheng Turng[2], Liqiang Yang[3], Adam Kramschuster[2], Junghoo Lee[2], and Srikanth Pilla[4]. (1) Departments of Mechanical Engineering & Materials, University of Wisconsin-Milwaukee, 3200 North Cramer Street, Milwaukee, WI 53211, (2) Department of Mechanical Engineering, University of Wisconsin-Madison, 1513 University Avenue, Madison, WI 53706-1572, (3) Department of Mechanical Engineering, University of Wisconsin-Milwaukee, 3200 North Cramer Street, Milwaukee, WI 53211, (4) Mechanical Engineering, University of Wisconsin-Milwaukee, 3200 North Cramer Street, Milwaukee, WI 53211

The effects of adding epoxy-based chain-extender on the properties of injection-molded solid and microcellular polylactide (PLA) were studied. PLA and PLA with 8 wt% chain-extender (PLA-CE) were melt-compounded using a twin-screw extruder. Solid and microcellular specimens were produced via a conventional and microcellular injection molding process, respectively. Various characterization techniques were applied to study the static and dynamic mechanical properties, cell morphology, and crystallization behavior. The addition of chain-extender reduced the cell size and increased the cell density in the microcellular specimens. It also decreased the crystallinity of PLA and increased the decomposition temperature. In addition, adding chain-extender slightly enhanced the average values of specific strength, specific toughness, and strain-at-break for both solid and microcellular PLA specimens.

395e Self-Assembled Soft Materials from Renewable Resources: a New Paradigm

George John, Department of Chemistry, The City College of New York,CUNY, Convent Avenue at 138th Street, New York, NY 10031

Our research efforts are deeply devoted towards developing building blocks from renewable resources to generate soft materials such as new surfactants, liquid crystals, lipid nanotubes and molecular gels. Present talk illustrates few successful examples of generating self-assembled soft materials from agri-sources, through simple organic transformations and by enzyme catalysis. To take these materials to the next level, we successfully showed the utility of these hydrogels as drug-delivery vehicles. Enzyme catalysis was used as a tool to make and break the hydrogels, which apparently triggered controlled drug-delivery. Intriguingly, by combining biocatalysis, with principles of green and supramolecular chemistry, we developed building blocks-to-assembled materials. Also address the advances that have led to the understanding of chiral behaviour and the subsequent ability to control the structure of glycolipid nanostructures-derived from renewable resources-and the resulting impact of this on future material applications. We foresee that these results will lead to efficient molecular design of supramolecular nanostructures and assemblies from plant/crop-based feedstocks, as new forms of materials, and energy needs.

395f Renewable (Truly Green) Highly Porous Nanocomposite Polymer Foams Synthesised Form Pickering - Emulsion Templates

Jonny J. Blaker, Xinxin Li, Angelika Menner, and **Alexander Bismarck**. Chemical Engineering, Imperial College London, South Kensington Campus, London, United Kingdom

Research efforts are being focused on the development of environmentally friendly bio-based nanocomposites in the desire to seek alternatives to petroleum based materials. In this work, novel bio-nanocomposite foams made from soybean-based resins and nanocellulose have been produced, which have potential in the manufacture of large composite parts, for example sandwich structures. Emulsion templating has emerged as an effective route to prepare porous polymers foams with a well-defined morphology since the latter is defined by the structure of the emulsion template at the gel-point of the polymerization.

Pickering emulsions are emulsions that are solely stabilised by small particles. These emulsions are extremely stable due to the irreversible adsorption of particles at the interface between the dispersed and continuous phase. Here we provide evidence that it is possible to stabilise Pickering medium and high internal phase emulsions (Pickering-M/HIPEs) of modified soybean oils having internal aqueous phase levels above 50% solely with hydrophobised silica and nanocellulose particles. Such emulsions can be used as emulsion templates for the synthesis of highly porous polymer foams, so-called poly-Pickering-M/HIPEs if the components of the continuous phase are polymerisable. Poly-Pickering-M/HIPEs are usually closed celled polymer foams, however, when using a green surfactant it is possible to synthesise open porous polymer foams with interconnected porous network structure.

We study the effect the internal phase level and amount of surfactant or particles used to stabilise the emulsions on the emulsion stability and droplet size and when polymerised on the foam morphology and mechanical properties.

Web Page: www3.imperial.ac.uk/portal/page/portallive/polymersandcompositesengineering/researchareas/biomaterials

459a Fundamental Thermo-Mechanical Property Modeling of Triglyceride-Based Thermosetting Resins

John J. La Scala, Army Research Laboratory, Building 4600, Aberdeen Proving Ground, MD 21005 and Richard P. Wool, University of Delaware, Composites Manufacturing Science Laboratory, University of Delaware, Newark, DE 19716.

Triglyceride-acrylates were prepared from various plant oils and model compounds using the unsaturation sites on the triglycerides. Unlike petroleum-derived resins, acrylated plant oils have a distribution of functionality ranging from 0 through 9 acrylates per triglyceride. This distribution was calculated using the distribution of unsaturation sites found on plant oils and then using successive binomial relationships to determine the probability of functionalizing a given functionality for a measured extent of functionalization. The cross-link densities of the resulting polymers were calculated using Flory-Stockmayer theory accounting for the extent of intramolecular cyclization. The glass transition temperature was clearly a function of the level of acrylation of triglyceride-based polymers and was modeled using simple empirical relationships. The glass transition temperatures of n-acrylated triglycerides were used as the relaxation modes for plant oil-based polymers. Using these relaxation modes, the Twinkling Fractal Theory of Tg was then used in conjunction with the calculated triglyceride distribution in order to predict the storage modulus, tan(δ), and loss modulus as a function of temperature. Essentially, the percentage of n-acrylated triglycerides with Tg less than that of the ambient temperature, the rubbery fraction, is the amount the modulus drops with temperature. The rubbery moduli for these polymers were calculated using the Theory of Rubber Elasticity and the calculated cross-link density. The tan(δ) was predicted based on the percentage of rubbery segments that were formed as the temperature was incremented from one relaxation mode to the next. The product of the storage modulus and tan(δ) is the loss modulus, and thus the loss modulus was also be predicted based on the polymer composition. There were some deviations between the experimental and predicted dynamic mechanical properties, but overall the model predictions matched the experimental results well. Nonetheless, this method offers an excellent method to approximate the dynamic mechanical properties as a function of temperature for a resin containing a distribution of n-functional cross-linkers.

459b Biobased Plastic Derived from Cereal Proteins

Dara L. Woerdeman, R&D Green Materials, LLC, 3624 Market Street, 5E (#542), Philadelphia, PA 19104

Biobased Plastic Derived from Cereal Proteins

Our work focuses on the development of an innovative processing method for yielding commercial-grade polymeric materials derived solely from biobased resources such as cereal proteins. Our approach for conditioning and processing the starting material leads to a solid plastic that is fully biodegradable. Our production method is expected to result in cost-competitive biobased polymer products with mechanical and performance attributes comparable to, or better than, those of their synthetic plastic counterparts. Plastic materials that are made entirely from cereal proteins are natural, and will, therefore, reduce pollution of the environment caused by toxic chemicals used in synthetic plastic manufacture and by non-degradable plastic waste accumulation.

459c Multifunctional Structural Green Nanocomposites from Biobased Thermoset Resin Systems: An Overview

Manju Misra, School of Engineering and the Department of Plant Agriculture, University of Guelph, Thornbrough Building, Guelph, ON N1G 2W1, Canada

Green/bio-based composites and nanocomposites are emerging as the premier group of materials for manufacturing and engineering applications in the 21st century. New thermoset biocomposites and nanocomposites have been developed in our laboratories and show tremendous promise for industrial applications like automotives, housing, aeronautical, other transportation systems, and marine industries, recreation equipments, farm equipments. These biocomposites and nanocomposites are processed from either hybrid biobased epoxy, biobased unsaturated polyester (USP) or bio-based polyurethane (PUR) matrices, and incorporate various natural fibers and/or E-glass fibers, carbon fibers, organically modified clays, untreated vapor grown carbon fibers (VGCF), fluorinated single wall carbon nanotubes (SWCNT), or surface treated alumina nanowhiskers as reinforcements. The biocomposites and nanocomposites were prepared by different processing techniques, such as sheet molding compounding (SMC), compression molding and vacuum assisted resin transfer molding (VARTM). All of the processed biocomposites and nanocomposites showed higher static and dynamic mechanical properties as compared to the neat matrix containing the same amount of functionalized vegetable oils. This presentation mostly highlights innovative value added non-food industrial applications for underutilized vegetable/plant derived oils, lignin and biomass derived biofibers.

Reinforcing Thermoplastics with Natural Fibers Derived from Biofuel Byproducts – Extended Abstract

Michael Fuqua & *Chad A. Ulven, PhD
Department of Mechanical Engineering and Applied Mechanics
North Dakota State University, Fargo, ND 58108
*contact: chad.ulven@ndsu.edu, 701.231.5641

Increasingly in the last few years, expansion seen in the biofuel sector has managed to dramatically change the dynamic of the United States' agricultural market. Crops traditionally grown as feedstock for consumption have found a lucrative market as fuel stock, which in turn has introduced new questions as to product allocation and usage. This maturing industry has brought with it a new source of lignocellulosic material in the form of biofuel production by-products, such as dried distillers grains with solubles (DDGS) from corn-ethanol. However, these by-products have a rather restricted application as livestock feed, due to their low nutritional value [1,2]. This causes their prices to be low relative to traditional lignocellulosic feed stock, and thus their potential use as reinforcement agents has become an attractive opportunity for production facilities with large quantities of derivative waste. With success being seen in the usage of materials traditionally considered filler agents without reinforcement capabilities, attention has begun to be placed on the usage of micro cellulose based natural fiber sources as well. It is in this pursuit that DDGS fibers have been explored as potential reinforcing fibers within polypropylene (PP).

The polypropylene used in this study was an isotactic PP trade named Formolene® 5100H supplied by the Formosa Plastics Corporation, USA, and the DDGS were supplied by Red Trail Energy, LLC, USA. The DDGS material was mechanically fractionated in a solution of the DDGS and 20 wt% acetone in a Fritsch Pulverisette 6 ball mill. The ball mill was run with 10mm stainless steel balls at 450 rpm for a total of 3 hours. This reduced the fiber size from multiple millimeters to approximately 20 μm colloidal particles.

In this work, PP/DDGS composite materials were explored using 10 wt% DDGS fibers within a PP matrix. Samples were prepared in order to determine the optimal processing methodology based on material performance. Mechanical investigation included both tensile and flexural tests performed according to ASTM standards D638 and D790, respectively. Water absorption tests were performed using a test modeled after ASTM D570. In this test, specimens were soaked in boiling water for 1 hour and tested for water absorption, after which they were tested in flexure to observe the resulting mechanical performance. Dynamic mechanical thermal analysis (DMTA) was also performed to observe changes in the glass transition temperature of the samples and subsequent crystallinity.

The hydrophobic nature of PP poses a potential problem in achieving good fiber/matrix adhesion in these systems however, as cellulose is an inherently hydrophilic polymer because of the numerous hydroxyl groups contained within it. To alleviate this obstacle, chemical compatibilizers have been developed which alter the surface of the hydrophilic cellulose fiber in order to improve interfacial adhesion between the fiber and matrix.

One such compatibilizer is maleic anhydride grafted polypropylene (MAPP), a waxy polymer system that has been proven useful in the processing and production of cellulose reinforced PP composites [3,4]. The MA group of the MAPP is able to react with the hydroxyl groups in cellulose, forming a bond between the MA group's carbon and the oxygen from the cellulose. At the same time, the PP portion of MAPP is able to co-crystallize with the unmodified PP, which allows for the MAPP to act as a bridge between the fiber and matrix [5,6]. MAPP is also able to compensate for insufficient breakup forces during processing, such as low shear stress, by reducing the interfacial tension between PP and the cellulose, which leads to finer dispersion of the fiber throughout the system [7]. MAPP was explored as a compatibilizer in both 5 wt% and 2.5 wt% concentrations with respect to the final composite, using Licomont® AR 504 supplied by Clariant, USA.

Another compatibilizer that was explored is organo-silane (R-Si-(OR')$_3$) [4]. Organo-silanes contain alkoxysilane groups which, when hydrolysized, are capable of reacting with surfaces having an abundance of OH molecules. As such, the coupling agent has the ability to close off the polar molecules in the cellulose fibers. This creates a hydrophobic coated fiber with a defined functional R group [8,9,10]. Although the functional group itself can not react with the PP matrix unless grafted during polymerization, the organo-silane does branch out and allows for mechanical interaction with the polymer chains of the matrix. Thus, mechanical interlocking can occur, and an improved fiber/matrix interface is created. In this work, two organo-silanes were used, an Aminopropyltriethoxysilane coupling agent supplied by Dow Corning®, trade name Z-6011 Silane, and an Aminoethylaminopropyltrimethoxysilane coupling agent supplied by Dow Corning®, trade name Z-6020 Silane. The Z-6020 Silane has the same defined functional H_2N group as Z-6011, however it is attached to a larger branched system, which brings with it the potential for better mechanical entanglement and interaction with PP chains.

For both the compatibilized and non-compatibilized fiber/PP blends, composite samples were produced by means of melt blending with a lab scale co-rotating twin screw extruder (Leistritz Micro-18/GL-40D). The screw rotation rate was set at 350 rpm and the temperature profile from hopper to die was as follows: 177°C, 182°C, 185°C, 189°C, 195°C, 195°C, and 195°C. This yielded a dwell time of approximately 30-40 seconds for the fiber/polymer system, minimizing prolonged exposure to degradation temperatures. Upon extrusion from the die, the extruded composite sample was quenched and pelletized. Sample pellets were then molded to shape for tensile and flexural specimens using injection molding. In this method, the composite pellets were injection molded into dog-bone tensile and flexural specimens using a Technoplas, Inc. SIM 5080 injection molder. The temperature profile from hopper to nozzle tip was as follows: 193°C, 204°C, 210°C, 216°C, and 204°C.

Mechanical testing demonstrated the importance of proper compatibilizer usage in obtaining optimal performance out of the DDGS fibers. As shown in Table 1, compatibilizer usage has significant trends on both tensile and flexural strength. First examining tensile strength, it is noted that the addition of 10wt% untreated DDGS caused a decrease in the tensile strength, and that this was not alleviated by treatment with either organo-silane treatment. However, both 2.5 and 5 wt% MAPP additions were shown to be beneficial in raising the tensile performance back to the base PP. Flexural strength performance shows a similar trend, with the MAPP compatibilization yielding improved strength results in comparison to neat fiber or silane treated.

From a polarity standpoint, all three compatibilizers should have performed similarly, since the MA group in the MAPP and the hydrolysised alkoxysilane in the organo-silanes are both acting as a bridge to de-polarize the hydroxyl groups in the cellulose components of the fibers. However, since chain entanglement of the PP and MAPP is virtually impossible [11], better fiber distribution is achievable in these composites than in the organo-silane treated ones. This in turn increases the likelihood of the fibers to carry load and improve the strength of the composites.

It was suspected that the lack of chain entanglement between PP and MAPP would cause fairly significant variation between the 2.5 wt% and 5 wt% MAPP concentrations. This is because if one concentration was in excess, the MAPP would be detrimental to adhesion because of the inability for entanglement between the modifier and neat PP. However, a concentration starved of enough groups to fully bridge the fibers would also perform poorly as part of the system would act as untreated fibers. Ultimately both loadings examined appeared to be within a window of optimal usage, in which there is not an extraneous excess of MAPP, but there is enough to create sufficient bridging between the DDGS and neat PP chains.

Table 1: Mechanical Properties of DDGS Reinforced Polypropylene Samples

Sample Type	Fiber Treatment	Elastic Modulus (GPa)	Tensile Strength (MPa)	Tangent Modulus (GPa)	Flexural Strength (MPa)
Neat PP	---	1.57 ± 0.09	27.6 ± 0.1	1.13 ± 0.04	37.3 ± 1.1
DDGS	Neat Fiber	1.66 ± 0.10	25.3 ± 0.2	1.30 ± 0.03	39.7 ± 0.3
	Silane Z-6011	1.55 ± 0.03	24.8 ± 0.1	1.30 ± 0.02	39.1 ± 0.4
	Silane Z-6020	1.59 ± 0.07	24.9 ± 0.1	1.29 ± 0.04	38.6 ± 0.7
	2.5 wt% MAPP	1.60 ± 0.03	27.9 ± 0.2	1.33 ± 0.03	41.6 ± 0.8
	5 wt% MAPP	1.59 ± 0.06	27.9 ± 0.2	1.34 ± 0.03	42.4 ± 0.6

Water absorption testing was performed in order to examine potential weatherability issues with the polymer systems while using the natural fiber reinforcement. As seen in Table 2, after one hour of immersion in boiling distilled water, the DDGS reinforced systems did not undergo significant changes in water weight absorption compared to neat PP. Previous research into water absorption of natural fiber reinforced polypropylene composites has demonstrated that significant absorption occurs over extended immersion [3]. However, this work has shown that under low immersion, mass gain from water absorption is not a significant factor.

Table 2: Water Absorption Results After 1 Hour Immersed in Boiling Water

Sample Type	Fiber Treatment	Weight Gain (%)
Neat PP	---	1.78 ± 0.34
DDGS	Neat Fiber	1.09 ± 0.42
	Silane Z-6011	1.50 ± 0.67
	Silane Z-6020	1.07 ± 0.72
	2.5 wt% MAPP	2.05 ± 1.05
	5 wt% MAPP	1.17 ± 0.63

Although there was a minimal observed change in water absorption by mass, the examination of the resultant flexural properties of the immersed composite samples demonstrates that water absorption did play a role on mechanical performance. Examining the flexural modulus, it can be seen in Figure 1a that water absorption caused a decrease in the tangential modulus of the yielded composites. However, it is worth noting that the samples which had been compatibilized saw a lessened degree of tangent modulus loss, indicating that the closing of the hydroxyl groups of the DDGS did play a role in preventing excessive water absorption. Flexural strength data, shown in Figure 1b, has a slightly different trend, in which MAPP usage showed to significantly prevent strength loss due to water absorption, while the organo-silanes and neat fiber did not. This would indicate that the MAPP creates more bonding with the open hydroxyl groups in the DDGS cellulose than the organo-silanes, a trend which is mirrored in the basic mechanical testing data.

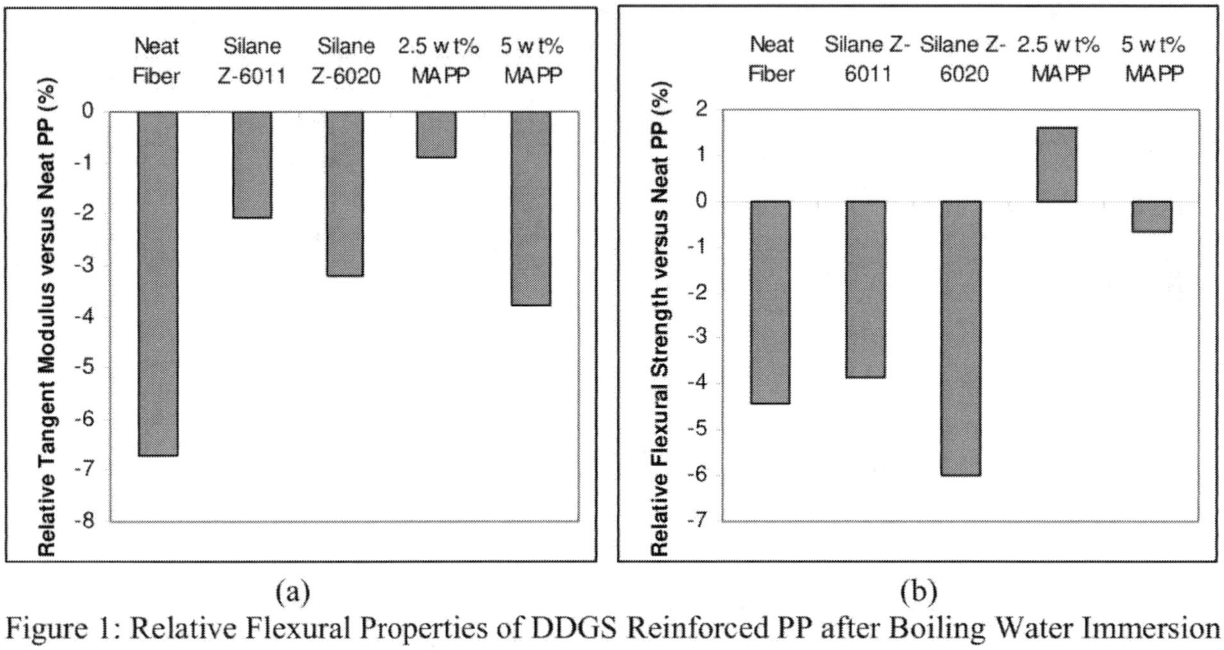

Figure 1: Relative Flexural Properties of DDGS Reinforced PP after Boiling Water Immersion

DMA work was performed using a TA Q800 DMA, operated at a frequency of 1 Hz, an amplitude of 20 μm, and a temperature ramp of 3°C/min. From this, analysis of the α-transition of the tan δ curve can be made, corresponding to the glass transition temperature (T_g) of the material as well as the crystallinity change in the specimens. As shown in Table 3, the addition of DDGS was found to decrease the glass transition temperature of the neat PP slightly. This indicates that DDGS fibers are being introduced between the polymer chains, increasing the spacing and free volume of the PP. However, such an interaction should also change the magnitude of the tan δ peak at T_g, which does not occur as shown in Table 3. As such, it is difficult to ascertain the level of polymer relaxation that was initiated due to the fibers. Previous work with natural fiber reinforced polypropylene has concluded that lignocellulosic natural fibers are too large to obstruct the mobilization of the PP macromolecular chains, and thus there is nothing to prevent the macromolecular segment from obtaining ordered alignment of the crystal lattice [12]. This would in turn prevent a crystallinity change, which falls in agreement with the tan δ peak value not changing.

Table 3: DMA Results

Sample Type	Fiber Treatment	T_g (°C)	tan δ
Neat PP	---	15.34	0.0575
DDGS	Neat Fiber	8.51	0.0592
	Silane Z-6011	10.10	0.0575
	Silane Z-6020	11.11	0.0570
	2.5 wt% MAPP	8.15	0.0575
	5 wt% MAPP	7.80	0.0600

[1] K.D. Rausch and R.L. Belyea. Appl. Biochem. Biotechnol. 128, 47 (2006)

[2] A. Batal and N. Dale. J. Appl. Poultry Res. 12, 400 (2003)

[3] W. Qui, F. Zhang, T. Endo, and T. Hirotsu. J. Appl. Polym. Sci. 87, 337 (2003)

[4] D.M. Panaitescu, D. Donescu, C. Bercu, D.M. Vuluga, M. Iorga, and M. Ghiurea. Polym. Eng. Sci. 47, 1228 (2007)

[5] T.A. Bullions, R.A. Gillespie, J. Price-O'Brien, and A.C. Loos. J. Appl. Polym. Sci. 92, 3771 (2004)

[6] J.A. Trejo-O'Reilly, J.Y. Cavaille, M. Paillet, A. Gandini, P. Herrera-Franco, and J. Cauich. Polym. Compos. 21, 65 (2000)

[7] K. Kim, H. Kim, and J. Lee. Polym. Eng. Sci. 41, 1963 (2001)

[8] X. Colom, F. Carrasco, P. Pages, and J. Canavate. Compos. Sci. Technol. 63, 161 (2003)

[9] M. Bengtsson and K. Oksman. Composites Part A 37, 752 (2006)

[10] J. Girones, J.A. Mendez, S. Boufi, F. Vilaseca, and P. Mutje. J. Appl. Polym. Sci. 103, 3706 (2007)

[11] L.W. Jang, E.S. Kim, H.S. Kim, and J. Yoon. J. Appl. Polym. Sci. 98, 1229 (2005)

[12] Fuqua, M.A., Ulven, C.A. J. Bio. Matls. and Bioen. 2, 258 (2008)

459e Biocomposite Materials for Building and Industrial Application

Satya Panigrahi, Department of Agricultural and Bioresource Engineering, University of Saskatchewan, 57 Campus Drive, S7N5A9, Saskatoon, SK S7N5A9, Canada

Agriculture is a major economy sector in Western Canada. Non-food agricultural fiber processing is an important sector in diversified agricultural industry. Oilseed flax, hemp and wheat are the crops grown and used as agricultural fibers in Saskatchewan. Especially oilseed flax and hemp crops can produce the strong bast fiber for commercial and industrial use. There are about 16,000 oilseed flax producers and flax crop had 785,000 hectares harvested area in 2006 (http://www.saskflax.com/). The potential salvageable oilseed flax straw on the Prairies is 500,000 to 1,000,000 tons annually. With flax fiber content of the straw ranging from 8% to 40%, this means that the potential pure fiber production from the oilseed flax presently grown on the Prairies would be between 100,000 and 250,000 tons annually. Flax and hemp bast fibers are the useful, strong and flexible raw material, because they are environmentally renewable, reusable and recyclable resource. Manufacturers use the bast fiber in plastic composites, insulation and a substitute for traditional wood products and synthetic fibers. Nowadays, a wide range of agricultural fiber products are already appearing on the market. Some of the products such as textile, paper pulp, biocomposite, counter tops, decks materials, pallets, construction materials, automobile parts and liners, rope, and insulation materials are made from flax and hemp fibers. However, there is still a problem in supplying consistent pure flax and hemp fiber at this moment. This hinders the growth of the biocomposite industries.

Development of biocomposite from hemp and flax fiber is still ongoing in the research sector. Different production processes of the biocomposite are being investigated. The biocomposite materials in extrusion, injection, compression and rotation molding grade are being investigated and developed to comply with the existing industrial process. Moreover, the utilization of natural fiber as supplement reinforcement for vacuum resin transfer molding (VRTM) is being introduced to the existing production in the fiberglass reinforced plastic industries. New techniques are being developed to manufacture the new products such as bioshingle, composite board and eco-block for the construction industries in Saskatchewan. The commercial egg trays is also manufactured from recycled materials including flax straw as fiber reinforcement in the processing

There is an increasing trend to make completely new types of composite materials used for building and industries by combining different resources. One of the biggest new areas of research in this field is combining natural fiber/straw with polymers. Fiber plants such as flax and hemp are increasingly used in manufacturing processes because consumers and businesses become more conscious of the environment and recognize the benefits of using renewable resources like natural fiber/straws. Canada produces the majority of flaxseed and hemp in the world. As such, a huge amount of flax and hemp straw is left in the growers' fields without suitable utilization. In our research, flax and hemp resource are used to make different products for construction, plastic, and automobile industry. Utilizing the developed technology to incorporate fiber/straw/shives into industry materials by different methods, some viable products are manufactured. The developed products include Eco-bricks, structural panel boards, bioshingles and green roof, oriented strand board used for new building system in all weather conditions (-40 to 40 ºC); and bio-based polymer, FlasticTM, HempsticTM and fibre reinforced polymer or recycled plastic composites used for plastics industries. The development of renewable, partially and completely biodegradable, environmentally friendly polymer and fiber-reinforced biocomposites from flax, hemp and other agricultural biomass and residues will reduce the Canadian dependence upon petroleum and other non-renewable resources. These products can provide a better or comparable strength than traditional materials. On the industrial side, a lot of manufacturing companies are looking at synthetic and natural resource for their own applications. The present lower cost and huge amount of flax and hemp resource are at an advantage to encourage construction, plastic and other industry manufacturers to shift their paradigm from 'tried and true' glass fiber and wood filler ingredients to renewable sustainable materials. Providing such sustainable materials will also reduce green house gas emission released from burning of plant straw or plastic land filling, and help achieve wild life preservation and the goal of eco-balance in Canada. It will create great opportunities for locally-based processing industries and create new local jobs, thus bring numerous economic benefits to society.

In Canada, the opportunity of bio-product industry is growing rapidly. The growth of biocomposite industry in natural fiber partly depends on the effective management in straw, retting and decortications from growers and producers. The researchers are doing their part to develop the new products, processes and technologies in biocomposite. However, the government is a major role to provide the funding to researchers, and to promote this innovative industry.

459f Development of Nanostructured Composites Via Spin-Assisted Layer-by-Layer Assembly

Ming Qin[1], Paul Podsiadlo[1], and Nicholas A. Kotov[2]. (1) Chemical Engineering, University of Michigan, 2300 Hayward St., 3074 H.H. Dow Building,, Ann Arbor, MI 48105, (2) Chemical Engineering Department, Materials Science Department, Biomedical Engineering Department, University of Michigan, 2300 Hayward, Ann Arbor, MI 48109

Natural composites possess exceptional mechanical properties. Of these, hybrid organic/inorganic composites are in particular well recognized and studied thanks to their excellent combination of strengths, hardness, and toughness. Some of the well known examples of such natural materials are seashell nacre, teeth, and bones. Nacre, which is composed of alternating layers of biopolymers and CaCO3 plates, has been especially well studied as a model material for preparation strong, tough, and flaw-tolerant composites. The uniqueness of nacre lies in its composition and overall mechanical properties: in spite of being composed of nearly 95% of brittle inorganic material and a few percent of weak biopolymers, the composite is 2x as strong and more than 1000x as tough as either one of the phases taken separately. These properties have yet to be reproduced.

Multilayered polymer-platelet nanocomposites have shown the ability to mimic the structure and mechanical properties of nacre, and as such they attracted much scientific attention recently. They also offer a potential for generation of materials with much greater mechanical properties than nacre since the nanoscale building blocks possess much higher properties when compared to their more coarse counterparts. However, the controlling factors which need to be addressed first are dispersion of the reinforcing platelets, effective load transfer at the organic/inorganic interface, and structural control. The layer-by-layer assembly (LBL) appears to address these problems, however slow preparation speed of traditional dipping method of LBL greatly inhibits its further application. In this proposal we propose, to develop a rapid, controllable spinning-assisted LBL method. This method will reduce the production time several orders of magnitude and it will potentially promote the application of LBL assembly.

459g Thermoforming of Hybrid Green Biocomposite from Bioplastic "Polyhydroxybutyrate-Co-Valerate", Wood Fiber and Talc

Sanjeev Singh[1], Amar K. Mohanty[2], Sher Paul Singh[1], Farhang Pourboghrat[3], and Manju Misra[4]. (1) School of Packaging, Michigan State University, 130 Packaging Building, MSU, East Lansing, MI 48824, (2) Department of Plant Agriculture & School of Engineering, Crop Science Building, University of Guelph, Guelph, ON N1G 2W1, Canada, (3) Mechanical Engineering Department, Michigan State University, 2241 Engineering Building, East Lansing, MI 48824, (4) School of Engineering and the Department of Plant Agriculture, University of Guelph, Thornbrough Building, Guelph, ON N1G 2W1, Canada

A fixed composition of polyhydroxybutyrate-co-valerate, maple wood fiber and talc were extruded and pelletized. A perform sheet out of these pellets were fabricated using compression molding technique and these sheets were thermoformed/ draped in to a dome shaped article. These articles were tested for their structural integrity.

537a Hot-Water Extraction and Forest Biorefinery

Thomas E. Amidon and Shije Liu. SUNY College of Environmental Science and Forestry, 1 Forestry Drive, Syracuse, NY 13210

Hot-water extraction is an efficient technique to separate hemicelluloses and extractives from woody biomass. In the absence of mineral acid or base, it is more environmentally beneficial for the recovery of chemicals from the extraction liquor. Membrane technology can applied for this purpose. The residual woody biomass has unique properties for improved solid products production.

537b Alkaline-Urea Pretreatment to Improve the Accessibility of Enzyme in Woodchips for the Biofuel Production

Yulin Deng and Yulin Zhao. School of chemical and biomolecular engineering, Georgia Institute of Technology, 500, 10th st. N.W., Atlanta, GA 30332

Alkaline pretreatment of spruce at low temperature in both presence and absence of urea was studied. It was found that the enzymatic hydrolysis rate and efficiency can be significantly improved by the pretreatment. At low temperature, the pretreatment chemicals, either NaOH alone or NaOH–urea mixture solution, can slightly remove lignin, hemicelluloses, and cellulose in the lignocellulosic materials, disrupt the connections between hemicelluloses, cellulose, and lignin, and alter the structure of treated biomass to make cellulose more accessible to hydrolysis enzymes. Moreover, the wood fiber bundles could be broken down to small and loose lignocellulosic particles by the chemical treatment. Therefore, the enzymatic hydrolysis efficiency of untreated mechanical fibers can also be remarkably enhanced by NaOH or NaOH/urea solution treatment. The results indicated that, for spruce, up to 70% glucose yield could be obtained for the cold temperature pretreatment (-15 °C) using 7% NaOH/12% urea solution, but only 20% and 24% glucose yields were obtained at temperatures of 23°C and 60°C, respectively, when other conditions remained the same. The best condition for the chemical pretreatment regarding this study was 3% NaOH/12% urea, and -15 °C. Over 60% glucose conversion was achieved upon this condition.

Web Page: www.chbe.gatech.edu/fac_staff/faculty/deng.php

537c Enzymatic Hydrolysis of Cellulytic Fibers Enhanced by Cationic Polyelectrolytes

John Reye and Sujit Banerjee. School of Chemical and Biomolecular Engineering, Georgia Institute of Technology, 500 10th Street, NW, Atlanta, GA 30332

"Enzymatic Hydrolysis of Cellulytic Fibers Enhanced by Cationic Polyelectrolytes"

Manufacturing pulp, paper and related products produces a considerable amount of cellulosic sludge. Approximately 5.5 million dry tons of sludge are disposed of each year by the pulp and paper mills, valued as a huge loss. Enzymatic hydrolysis of cellulosic waste followed by anaerobic conversion to methane may provide a viable alternative to waste disposal methods such as incineration and landfilling. This is based on our recent finding that cationic polymers increase the enzymatic hydrolysis rate of two model substrates: cellulosic fiber and cornstarch. Preliminary findings show a 20% or greater increase in initial rate of hydrolysis for appropriate systems. At high concentration (1000 ppm) c-PAM is advantageous for cellulosic fiber, while lower concentration (100 ppm) more positively enhances cornstarch hydrolysis. This concept may result in a greater breakdown of cellulosic solid waste, which could be applicable to a variety of other systems where an enzyme is breaking down an insoluble substrate.

537d Chemical Analysis and Kinetics of Wood Extract Hydrolysis

Shije Liu[1], David Kiemle[2], Alan Shupe[3], Ruofei Hu[3], and Yang Wang[3]. (1) SUNY College of Environmental Science and Forestry, 1 Forestry Drive, Syracuse, NY 13210, (2) Chemistry, SUNY College of Environmental Science and Forestry, 1 Forestry Drive, Syracuse, NY 13210, (3) Paper and Bioprocess Engineering, SUNY College of Environmental Science and Forestry, 1 Forestry Drive, Syracuse, NY 13210

H NMR and C NMR are applied in combination to study the kinetics of wood extract hydrolysis. Because of the complexity of the wood extracts, single component NMR is not enough for the detection of both monomeric sugars and oligomers. This study attempts to establish a standard for the biomass extract analysis.

537e Production of Ethanol from Forest Residues

Kristina Knutson, Arthur Ragauskas, W. J. Frederick Jr., and Kristiina Iisa. Georgia Institute of Technology, Atlanta, GA 30332

The value of wood waste can be enhanced considerably above its fuel value by extracting carbohydrates from it before burning it, and converting the carbohydrates to fuel-grade ethanol. The process conditions for extracting and fermenting carbohydrates from various southern pine residues (bark, needles, twigs, and branches) were evaluated, and the characteristics of the post-extracted wood waste determined. An economic assessment of incorporating forest residue and hog fuel extraction and fermentation in a pulp mill is included.

537f Effect of Surfactants, Flocculants and Hydrolyzing Method on Centrate Recovery from Whole Stillage in Corn Bio-Ethanol Plant

Mohammadali Safavieh, Marian Simo, and Vladimir Hlavacek. Chemical and Biological Engineering, University at Buffalo, 303 Furnas Hall, Buffalo, NY 14260

This poster is about the effect of adding the different surfactant and flocculants on separation of water of whole stillage in centrifuging part of ethanol plan. Basically we used different concentrations of Commercial detergent and some non-commercial surfactants such as Poly ethylene glycol 400 and Vinyl acetate. Among different surfactants that we used, Poly ethylene glycol 400 (PEG 400) has a great effect on centrate recovery and separation of water from whole stillage. Alsowe used different flocculants. Combination of "307 Tramfloc" flucculant and commercial detergent can give us good result too. We analyzed effect of ultrasound on whole stillage recovery. Ultrasound does not have effective effect on separation of water from whole stillage. Also we hydrolyzed the whole stillage under different conditions of pressure and temperatures.By hydrolyzing of whole stillage at high pressure (420 Psi) we can increase the efficiency of centrate recovery.

Key words: Whole Stillage ,Surfactant ,Flocculants, Hydrolyze

623a Renewable Fuels Combustion Technologies

Robert S. Morrow, Amy Malko, and Tom Giaier. Detroit Stoker Company, 1510 E. First Street, Monroe, MI 48161

For six decades, Detroit Stoker Company has been designing and supplying combustion systems for a wide variety of biomass fired applications. With current public concerns regarding global warming and Green House gases, these combustion systems and fuels utilized have become more sophisticated. This paper will present experiences of not only North American but also European facilities with examples of emission performance and ability to provide equipment for a wide range of fuel types, fuel moistures and opportunity fuel generated from process areas and agricultural waste streams. In addition, use of computational fluid dynamics (CFD) modeling and results from pilot scale combustion testing will be discussed and how these tools assisted in design and supply of these commercial combustion systems.

623b Biomass Gasification in a Stoker Gasifier

Kevin Whitty, Institute for Clean and Secure Energy (ICSE), Chem. Eng. Dept., University of Utah, 50 South Central Campus Drive, Room 3290 MEB, Salt Lake City, UT 84112, Eric G. Eddings, Dept. of Chemical Engineering, University of Utah, 50 So. Central Campus Drive, 3290 MEB, Salt Lake City, UT 84112-9203, and Robert S. Morrow, Detroit Stoker Company, Monroe, MI 48161.

As a consequence of increasing energy costs and a push towards carbon-neutral energy sources, the forest products industry is interested in minimizing its use of fossil fuels and increasing energy production from biomass. One solution is to gasify biomass to create a combustible gas that could be used to displace natural gas or oil in boilers and lime kilns. The biomass gasifier should be simple and robust, with low capital and operating expenses. Detroit Stoker Company has over 100 years experience with biomass combustion and has many stokers installed at forest products facilities throughout the world. If a stoker is operated under reducing conditions (stoichiometric ratio < 1.0), it effectively becomes a gasifier and could conceivably offer a low cost gasification solution for the forest products industry. In the study presented here, Detroit Stoker Company and the University of Utah performed a first-of-a-kind gasification campaign using a 300 kWth pilot stoker facility at the University of Utah. Results indicate that a stoker-type gasifier can produce a quality producer gas (synthesis gas) while maintaining good heat release from the bed. In this presentation, overall performance of a stoker gasifier is discussed and data relating to system performance and product gas quality is presented.

623c Biorefinopttm. A Practical Tool for Process Design and Economic Analysis of Integrated Biorefineries

W. J. Frederick and Steven Lien. Georgia Institute of Technology, Atlanta, GA 30332

BioRefinOptTM is a software tool designed specifically for preliminary design and economic analysis for integrated biorefineries. In this context, an integrated biorefinery is an integrated industrial complex that contains at least a chemical pulp mill, a bio-based fuels plant, and a combined heat and pwer (CHP) plant. BioRefinOptTM is capable of simulating an alkaline pulp mill integrated with an ethanol plant that utilizes wood as feedstock. It has been used in several evaluations of industrial biorefinery configurations. Two specific applications have recently been reported (1,2).

BioRefinOptTM performs mass and energy balances around an entire biorefinery complex. It assesses the capacity rating for each unit operation in the integrated biorefinery complex, and it accounts for changes in throughput caused by process changes resulting from integration of an ethanol plant with a pulp mill. The input data and worksheets within BioRefinOptTM allow it to be modified to any alkaline pulp mill and/or ethanol plant configuration.

This paper describes the conceptual design of BioRefinOptTM and presents details of the models that comprise it. Examples of applications are presented.

References:

1. Frederick, W.J. Jr., Lien, S.J., Courchene, C.E., DeMartini, N.A., Ragauskas, A.J., Iisa, K. Co-Production of Ethanol and Cellulose Fiber from Southern Pine: A Technical and Economic Assessment. Accepted for publication in Biomass & Bioenergy, Oct 10, 2007.

2. Frederick, W.J. Jr., Lien, S.J., Courchene, C.E., DeMartini, N.A., Ragauskas, A.J., Iisa, K. Production of Ethanol from Carbohydrates from Loblolly Pine: A Technical and Economic Assessment. Bioresource Technology 99 (2008) 5051–5057.

David J. Krygsveld, Entropy Associates Inc.
Thursday, November 20, 2008: 8:30 AM-11:00 AM
Room 204-A (Pennsylvania Convention Center)

Forest and Plant Bioproducts Division (17)

#623 - Topics in Energy from Forest Bioproducts: Economics, Life Cycle Analysis, and Improved Thermal Processes (17004)
Overview: A variety of forest bioproducts topics will be considered, including the engineering and economics of biomass energy systems, lifecycle analysis of forest bioproducts and biofuels, and improvements in processes that utilize energy from biomass for power generation and process heat.

Chair:	Kenneth Nichols	
CoChair:	Bandaru V. Ramarao	
8:30 AM	623a	Renewable Fuels Combustion Technologies **Robert S. Morrow**, Amy Malko, Tom Giaier
9:00 AM	623b	Biomass Gasification in a Stoker Gasifier **Kevin Whitty**, Eric G. Eddings, Robert S. Morrow
9:30 AM	623c	Biorefinopttm. A Practical Tool for Process Design and Economic Analysis of Integrated Biorefineries **W. J. Frederick**, Steven Lien
*10:00 AM	623d	Experience with Ash Modifiers to Control Pluggage of Biomass Furnaces and Heat Transfer Surfaces **David J. Krygsveld**
10:30 AM	623e	Economic Analysis of Optimal Biomass Utilization in An Integrated Biorefinery **Norman E. Sammons Jr.**, Wei Yuan, Susilpa Bommareddy, Mario R. Eden, Burak Aksoy, Harry T. Cullinan

*10:00 AM
623d

Experience with Ash Modifiers to Control Pluggage of Biomass Furnaces and Heat Transfer Surfaces

David J. Krygsveld, Entropy Associates Inc., 266 Millar Avenue, Miramichi, NB E1V 3B6, Canada
Title
Experience with ash modifiers to control pluggage of biomass furnaces and heat transfer surfaces
Review of stages of combustion.
Types of ash.
Source of problem ash.
Correcting problem.

David J. Krygsveld, Entropy Associates Inc.
266 Millar Avenue
Miramichi
New Brunswick E1V 3B6
Canada

Experience with ash modifiers to control pluggage of biomass furnaces and heat transfer surfaces

David J. Krygsveld, Entropy Associates Inc.

Introduction

As a field engineer, I interface with operations personnel of varying background who, of course, have direct experience with their furnaces and are experiencing plugging due to ash build-up in the furnace and downstream heat exchange surfaces.
In order to ensure that these people understand the process from the same point of view, I explain as follows.

David J. Krygsveld, Entropy Associates Inc.
266 Millar Avenue
Miramichi
New Brunswick E1V 3B6
Canada

Review of stages of combustion.

By definition, all biomass units burn carbon.
Naturally each furnace operator will claim that his fuel is unique in some way and so his problem is different. This is not relevant to understanding ash formation.
Biomass is a carbohydrate with moisture and soil contaminants either mixed or bonded in its structure.
In order to burn, the material is introduced to the furnace, where it begins to heat up to 212° F, the vaporizing temperature of water. In this heating process, any contaminants with a lower vaporizing temperature, such as alcohol and turpentine, will evaporate at their boiling point and join the fume from the bed, perhaps combusting.
The fuel remains at the boiling temperature of water until all the moisture is gone before the temperature increases again, this time rising through the volatile temperatures of entrained salts. When the ignition temperature of carbon is achieved at about 600 ° F, the carbohydrates will decompose and the carbon will burn to carbon monoxide, provided that the air supply is restricted. The carbon monoxide will rise from the initial drying/burning zone until it is supplied with a secondary supply of air to burn carbon monoxide to carbon dioxide at about 1100 ° F.
The carbon monoxide to carbon dioxide flame is relatively unaffected by water and salts the way the drying/burning zone

David J. Krygsveld, Entropy Associates Inc.
266 Millar Avenue
Miramichi
New Brunswick E1V 3B6
Canada

is. Although the average flame temperature in this zone is measured as less than 1900 °F, the fact that glass is melted in this zone suggests that a temperature of over 2500 °F is actually achieved in some area. This is likely due to combustion of hydrogen and other high flame temperature materials.

Types of ash.

Ash in biomass is of three main types; cold, hot and melted. Cold ash that remains cold usually falls to the grate and is removed by equipment designed for this and so is not a problem. Hot ash, that is ash that is not melted, carries with the main flow of flue gas and is removed from the downstream zones by ash removal equipment and poses no problem because it is not sticky.

Melted ash, usually a very small fraction, will flow with the flue gas and stick to the first surface that cools it to below its fusion point and here lies the problem.

David J. Krygsveld, Entropy Associates Inc.
266 Millar Avenue
Miramichi
New Brunswick E1V 3B6
Canada

Source of problem ash.

Melted ash, or ash which has changed state due to high temperature, is the source of problem pluggage.

Because most biomass entrains some sand in its cells, it is impossible to avoid ash formation in most furnaces, due to the areas where the flame temperature is over the melting point of glass.

Correcting problem.

The strategy concentrates on removal of ash only.

There are two principal strategies to cope with ash formation; design of the furnace and removal of the ash.

Since each user has an already built unit, re-design is not usually an option.

Since the ash is glassy, it is very hard, although brittle and cannot be removed by anything short of mechanical impact such as jackhammering.

Chemistry can be applied to change the crystal strength of the ash, making it friable and easier to remove.

This chemistry is best applied in the carbon monoxide flame where the glass is melted and the chosen ingredient can melt with it.

Several companies supply materials of this type that can often be used to financial advantage.

David J. Krygsveld, Entropy Associates Inc.
266 Millar Avenue
Miramichi
New Brunswick E1V 3B6
Canada

In my experience, operators often ask for guarantees. The chemical additive is guaranteed to alter ash characteristics and it is at this point that the operator needs to take advantage of this alteration to improve performance.

Question period.

David J. Krygsveld, Entropy Associates Inc.
266 Millar Avenue
Miramichi
New Brunswick E1V 3B6
Canada

623e Economic Analysis of Optimal Biomass Utilization in An Integrated Biorefinery

Norman E. Sammons Jr.[1], Wei Yuan[1], Susilpa Bommareddy[1], Mario R. Eden[1], Burak Aksoy[2], and Harry T. Cullinan[2]. (1) Department of Chemical Engineering, Auburn University, 230 Ross Hall, Auburn University, AL 36849-5127, (2) Alabama Center for Paper and Bioresource Engineering, 242 Ross Hall, Auburn University, AL 36849

The integrated biorefinery has the opportunity to provide a strong, self-dependent, sustainable alternative for the production of bulk and fine chemicals from polymers, fiber composites and pharmaceuticals to energy, liquid fuels and hydrogen. With such a wide range of processing steps and possible products, it is obvious that identification of the optimum process structure can not be done based on heuristics or rules of thumb. Thus, there is a need for systematic, reliable methods capable of incorporating different levels of process detail in the decision making framework. By constructing simulation models for various biorefining processes, determining variable and fixed cost components of these processes, and combining this cost data with information on expected return and market conditions, firms interested in biorefining will be able to determine the effects of changes in value as a result of deciding which products to pursue among the many biorefinery possibilities. Through quantifying the level of environmental impact using EPA's WAR algorithm, sustainability could then be measured on a relative scale. An inherent benefit of the proposed framework comes from the decoupling of the complex models from the selection step, which results in the ability to adapt to new developments within any of the processing steps and thus also incorporate novel innovative production processes in the decision-making framework. In this way, experimental and theoretical efforts can supplement each other in a synergistic manner, by providing direction and data for continued work. Changing market conditions may dictate a dynamic optimum for the allocation of resources and production capacity, resulting in a myriad of possible long-term product portfolios and a need for a net present value perspective that takes into account the time value of money. This contribution will illustrate the strategy for developing the decision making framework as well as highlighting the flexibility of the framework to utilize data from technological breakthroughs in the field of biorefining.

EXPERIMENTAL AND NUMERICAL INVESTIGATION OF WOOD DRYING PROCESS USING MODIFIED INFRA-RED HEATING TREATMENT

Erzsébet Cserta, Department of Metallurgy, Chair of Thermal Processing Technology, University of Leoben, Leoben, Austria
Dr. László Könözsy, Department of Metallurgy, Chair for Simulation and Modelling of Metallurgical Processes, University of Leoben, Leoben, Austria
Gergely Hegedűs, Kentech Ltd., Hungary
Prof. Harald Raupenstrauch, Department of Metallurgy, Chair of Thermal Processing Technology, University of Leoben, Leoben, Austria

Introduction

The wood plays an important role as a natural material in the industry. For the engineering practice, the selection of the adequate drying method is indispensable to produce high material quality of the product. Depending on the original material structure and the climate conditions, several drying methods are used according to the type of the energy transfer into the sample. The drying process based on the heat radiation became more and more important research field in the past decades. Essentially, all these technologies use electromagnetic waves to heat up the wood [1-6]. The drying of organic raw materials using infra-red (IR) radiation also takes place among these processes as a relatively new technology.

The wood drying process can be defined as a simultaneous heat and moisture transport with local thermodynamic equilibrium at each point within the probe. The process can accuratly be determined by knowing the temperature distribution in time and space. The complex internal structure of the wood causes difficulties in the simulation at microscopic level. Therefore, the most classical models use homogenization [7-9] or space averaging [10] techniques.

The present paper deals with the drying process of spruce wood *(Picea abies)* in a newly developed IR drying furnace according to the patent Nr. PO 700 453. The required frequency field is ensured by heating wires using modificator layers. During the process, the cellulose molecules are depolymerized, therefore, the water can easily exit from the wood. Several wood types were analysed at different drying conditions to determine the optimal heating parameters. The results of spruce wood was chosen to show the efficiency of the technology. The experimental investigation of the process can be found in the first section.

For the better understanding of the proposed IR drying process, a one-dimensional model has been set up to investigate numerically the heat transfer in wood slab. In the second section of this work, the finite difference method has been used to obtain an approximate solution of the temperature profile. The temperature conductivity has been taken into account based on the experimental data system. The numerically predicted results have been compared to temperature measurements and good agreements with experimental data were obtained.

Experimental investigation

Experimental setup

A newly designed IR furnace has been set up to investigate experimentally the drying process. The flow sheet of the system is shown in the Fig. 1.

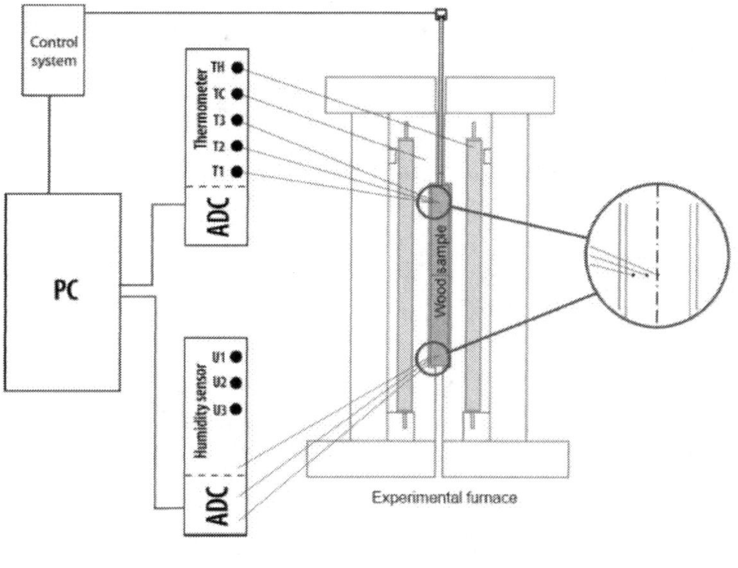

Figure 1. The flow sheet of the experimental set-up; a) the IR furnace; b) the regulation system; c) the data acquisition system.

The test facility consists of the drying furnace with the IR heating system, the data acquisition system and the regulation system. The size of the drying area can be controlled according to the sample size by altering the distance between the heating surfaces. The heating surface is built up of 12 heating wires using modificator layers at the two vertical walls around the sample. The heating power of the wires is completely 4 kW. The infra-red radiation was generated at the wavelength between 700-1000 μm. The regulation system is designed to be eligible both manual and automatic controlling. A digital adquisition system was worked out to store the whole data set. The equipment enables simultaneous measurement of the temperature distribution and moisture content. The temperature was measured using Dixell and Lea K-type thermo-elements. The temperature values appear on the digital screen of the measuring meter and a time-programmed camera is used for recording the data. There were test holes drilled in the drying wood where the sensors of the temperature detectors were located. The moisture content was measured in the test holes as well, and the values were obtained by using Elbez WHT 860 moisture sensors.

Samples

The investigated spruce wood samples were green or air-seasoned timbers. The green ones had an average initial moisture content which was higher than 40%. Therefore, the initial moisture content of the seasoned samples was dependent on the level of the pre-drying process. The final moisture content was under 8% in the overall measurement series. The average initial sample size is a 50 x 100 [mm/mm] cross section. The technological parameters and the physical properties of the spruce samples can be found in Table 1.

Table 1. Technological parateres and physical properties of the spruce samples

Wood sample type	Total heating time [min]	Incubation time [min]	Initinal moisture content [%]	Final moisture content [%]	Average moister loss [%]	Heating wire temperature [°C]
TF7	330	90	45-50	7>	40-42	340
TF8	360	110	42-64	6-25	36-39	300
TF12	1890	1700	44	6-8	37	200
TF11	500	110	18-33	8>	15-25	300
TF105	180	10	18-21	6>	12-15	480

The physical process of wood drying with IR heating treatment

The IR wood drying technology was devided into two parts according to the location of the wood sample. The wood is installed into the drying furnace in the first part of the process, which is the IR furnace drying process. The drying is caused mainly due to heat radiation. The heat transfer occurs due to conduction, when the wood is taken out of the oven. The drying continuous on the free air in the room temperature, in this way, the second part of the process is the post-drying process.

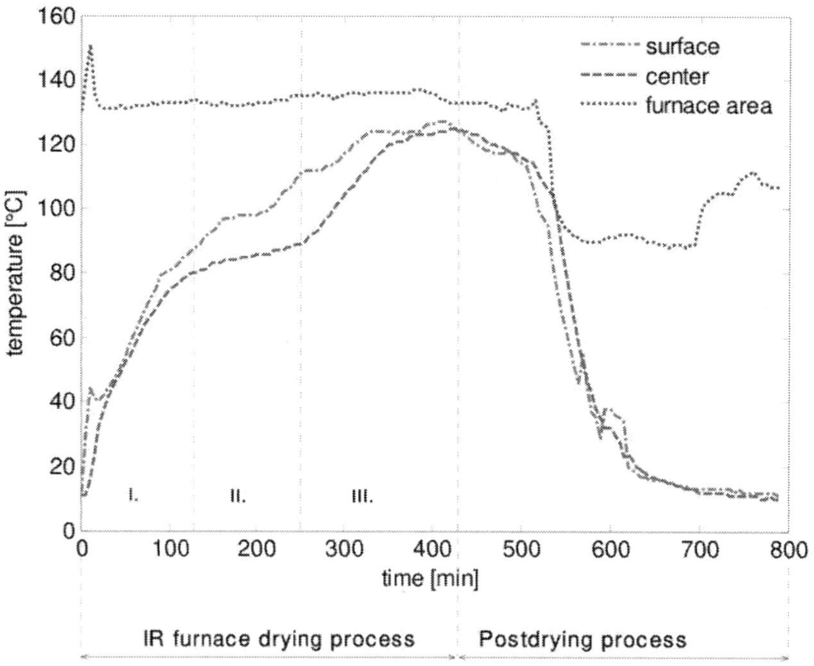

Figure 2. The surface and the center temperature of the TF11 spruce timber depending on time using infra-red radiation, and the average temperature of the furnace area

Three drying periods were determined depending on the temperature distribution in the sample. These are the heating-up etap (I), the incubation etap (II) and the reverse heating etap (III) (see Fig. 2.). The first drying time interval was considered as the heating-up process. The temperature of the sample surface increases faster than the temperature of the wood center. The increment of the core temperature is stopped at around 80 ^{o}C, and the second drying time interval starts at that point. The core temperature profile shows a stagnation in this significant time interval. When the incubation point is achieved, then an evaporization front startes to move from the surface towards to the center. This phenomenon refers to an internal boiling of the water, and the introduced heat is transformed to latent heat. Corresponding to the low temperature at the stagnation interval, a local suction effect is produced behind the evaporization front. The low pressure is assumed to be the effect of the osmotic mechanism. The semipermeable membrane is built up by cellulose mulecules, which let through only the water molecules. The larger salt molecules are blocked, therefore a concentration difference occurs between the two sides of the cellulosic membrane. The internal osmotic effect is well-balanced by the water-vapor pressure. When the evaporative front reaches the axis of the sample, the core temperature increases faster than the surface temperature. The core temperature can even be higher than the surface temperature. This is an opposite phenomenon compared to the convective drying process. The evaporative front can achive the core, because the cellulose membrane has been depolymerized. The produced temperature gradient helps the moisture content to exit easily, when the core temperature is higher than the surface temperature. When the core temperature reaches or will be above the surface temperature, then the sample will be taken out of the oven. The post-drying process goes further on until the core temperature will be equal to the room temperature.

Brief description of the mathematical model and the numerical solution

A one-dimensional model has been set up to investigate numerically the heat transfer in the wood slab (see Fig. 3.). The finite difference method has been used to obtain an approximate solution of the temperature profile using explicit and implicit schemes. The temperature conductivity, which is one of the most important characteristic of the material, has been taken into account based on the experimental data system. The numerically predicted results have been compared with temperature measurements concerning different process parameters. Considering the complex structural inhomogenity of the wood, it is necessary to have simplifying assumtions. For this reason, the wood is considered as a homogeneous material, the convective heat transport is not taken into account, the moisture content behind the evaporative front is equal to the initial moisture content, and there is local thermal equilibrium inside the material.

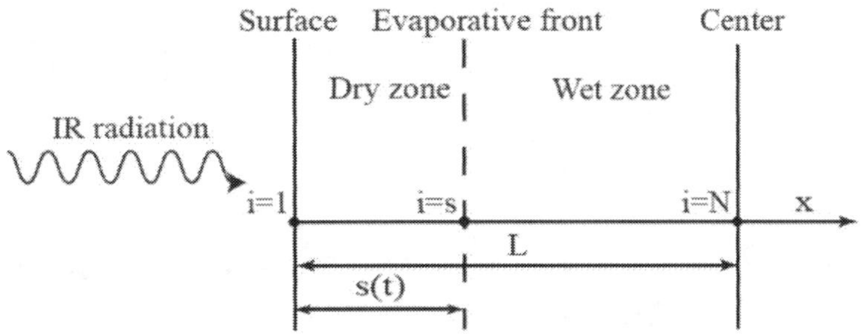

Figure 3. One-dimensional heat conduction model taking into account the movement of the evaporative front during the drying process

The movement of the evaporative front is modelled and it is considered as the moving boundary between the dry and wet part. The Dirichlet boundary conditions at the corners are time-dependent based on the measurements. The radiative term and the latent heat referring to the phase change along the evaporative front is included in the Neumann boundary condition at the moving boundary. The heat conduction equation is solved for both of the dry and wet part of the wood, therefore the governing equation for the dry side of the sample is

$$\frac{\partial \vartheta_1}{\partial t} = \frac{\partial}{\partial x} \cdot \left[\kappa_1(\vartheta_1) \cdot \frac{\partial \vartheta_1}{\partial x} \right], \tag{1}$$

and for the wet side of the sampe is

$$\frac{\partial \vartheta_2}{\partial t} = \frac{\partial}{\partial x} \cdot \left[\kappa_2(\vartheta) \cdot \frac{\partial \vartheta_2}{\partial x} \right], \tag{2}$$

where $\kappa_1(\vartheta_1)$ and $\kappa_2(\vartheta_2)$ are the temperature conductivities of the dry and wet zones of the sample corresponding to ϑ_1 and ϑ_2 temperatures. The temperature dependence of the temparature conductivity for the dry side of the wood is the sum of the conductive term

$$\begin{aligned} \kappa_{cond}(\vartheta_1) = & -1.3185 \cdot 10^{-2} + 1.723 \cdot 10^{-7} \vartheta_1 - 1.27 \cdot 10^{-8} \vartheta_1^2 + 7.8071 \cdot 10^{-11} \vartheta_1^3 - \\ & - 6.913 \cdot 10^{-13} \vartheta_1^4 + 1.055 \cdot 10^{-14} \vartheta_1^5, \end{aligned} \tag{3}$$

and the radiative term

$$\kappa_{rad}(\vartheta_1) = 17.15 \cdot \varepsilon_{rad} \cdot \sigma \cdot \vartheta_1^3 \cdot d, \tag{4}$$

where ε_{rad} is the radiation emissivity of the wood, σ is the Stefan-Boltzmann constant, and d is the equivalent pore diameter [11]. The radiation emissivity can be chosen according to

$$1 \geq \varepsilon_{rad} \geq 0.9. \tag{5}$$

The surrounding air and the wood can be considered magnetically transparent at the modified frequency field. The incident radiation, emitted from the heating wires, reaches the wood and penetrates into the inhomogenouos structure until it is absorbed by the moisture of the wood. Consequently, the most important absorbtion effect is evolved in the evaporative front. This plane of the evaporization is taken into account as an internal wet surface of the wood, where the IR radiation is absorbed. Behind this front, the effect of the radiative heat term can be neglected in the wet zone of the sample. According to Eqs. (3)-(4), the following expression may valid for the temperature conductivity

$$\kappa_1(\vartheta_1) = \kappa_{cond}(\vartheta_1) + \kappa_{rad}(\vartheta_1). \tag{6}$$

The temperature conductivity $\kappa_2(\vartheta_2)$ is considered as a constant in the wet side of the sample. Since, the movement of the evaporative front is computed, therefore it is necessary to specify boundary condition at the moving boundary. Concerning the above mentioned assumptions, the moving boundary condition of the evaporative front can be fullfilled as

$$\alpha_1(\vartheta_1) \cdot \left. \frac{\partial \vartheta_1}{\partial x} \right|_{s(t)} = \alpha_2(\vartheta_2) \cdot \left. \frac{\partial \vartheta_2}{\partial x} \right|_{s(t)} + S_{ph} + S_{rad}, \tag{7}$$

where

$$\begin{aligned} \alpha_1(\vartheta_1) = & -9.23 \cdot 10^{-3} + 1.213 \cdot 10^{-7} \vartheta_1 - 8.858 \cdot 10^{-9} \vartheta_1^2 + 5.115 \cdot 10^{-11} \vartheta_1^3 - \\ & - 4.839 \cdot 10^{-13} \vartheta_1^4 + 7.386 \cdot 10^{-15} \vartheta_1^5 \end{aligned} \tag{8}$$

is the thermal conductivity of the dry section of the wood, S_{ph} is the latent heat corresponding to the phase change during the evaporization, and S_{rad} is the additional heat term from the radiative heat absorbtion at the evaporative front. The S_{ph} is given according to [12,13] considering the movement of the phase change as a function of time. The value of the S_{rad} is computed from the mean radiation power density locally absorbed by the sample [5,6]. The thermal conductivity $\alpha_2(\vartheta_2)$ is considered as a constant in the model. The parameter values are implemented in the simulation based on the measurements.

The solution of Eqs. (1)-(2) time-dependent and non-linear parabolic heat conductions were obtained by explicit finite difference method. An in-house MATLAB code was implemented to estimate the temperature distribution in the wood slab. The temperature dependence of the temparature conductivity were taken into account based on the Eq. (3) in the dry side of the sample. The choise of the temporal and spacial terms are not arbitrary in the explicit method. It is essential to determine these terms according to the accuracy and stability conditions. The stability of the algorithm has to be ensured, therefore the following formula can be derived for the time-step size to satisfy the stability criterion as

$$\Delta t \leq \frac{(\Delta x)^2}{2\kappa_1(\vartheta_1)}. \qquad (9)$$

For solving the one-dimensional non-linear parabolic heat conduction equation, the explicit method is adviced, because it can be easily implemented. For solving the Eqs. (1)-(2) using an implicit method, it is necessary to linearize the right hand side of the equations. The Crank-Nicolson implicit scheme was used. This method computes temperature values between two time-levels. For implicit solution of the one-dimensional heat conduction problem, a tridiagonal system of linear equations has to be solved. The numerically predicted results will be discussed in the next section.

Results and discussion

For reaching the required material quality in the wood industry, the optimal heating parameters is to be determined. The heating rate was altered and adjusted according to the pre-drying level of the samples. These samples were experimentally investigated with different initial moisture content (see Table 1.). The decomposition time of the cellulose during the furnace drying process is strongly determined by the initial stage of the wood. The air-seasoning process has strong effect on the drying. When the cellulosic membrane is already decomposed during this process, afterwards the incubation period is insignificant. An inflection point can only be detected as it is shown in the Fig. 4.c. The initial free moisture content has small influence on the heating time, because the major part of the installed heat is used for the elimination of the chemically bound water (see Fig. 4.b.). After the decomposition of the cellulosic membrane, the moisture can easily exit without causing internal cracks. It is also essential to take care of the internal temperature gradient during the incubation time. When the internal temperature difference is kept under 20 °C degree, the internal crack formation can be avoided. Above this temperature gradient value, the cells may be exploded and internal cracks might be formed in the slab (see Fig. 4.d.). It is important to find the optimal heating rate, because the core temperature can reach the surface temperature only in that case, when the radiation intensity is high enough (see Fig. 4.a.).

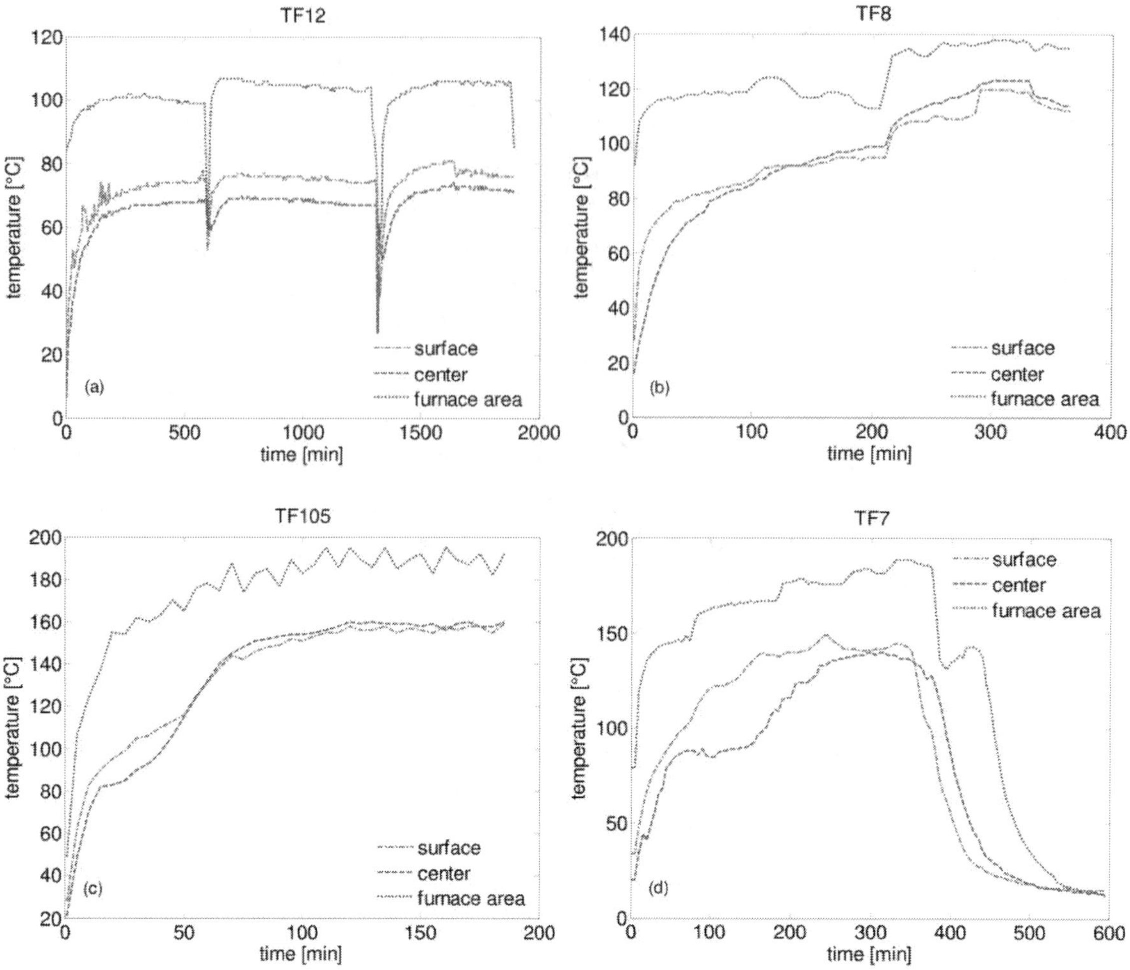

Figure 4. The surface and the center temperature of green (a, b, d) and seasoned (c) spruce timbers depending on time, and the average temperature of the furnace area

The measured temperature curves in the Fig. 4.a-d. give the boundary values for the mathematical model. The temperature were measured at the surface, the core and the mid-point of the computational domain. In this way, it was possible to test the model using time-dependent boundary conditions. The measurements data were implemeted, and the aim of the numerical investigation was to analyse how the computed curves approch the measured points. The computed curves in the domain were displayed in every 20th minutes (see Fig. 5 and Fig. 6.). The predicted results were compared to the experimental data. It means that the mathematical model can reproduce the measured temperature distribution. The simulation curves fit to the measured point (see Fig 5.a-b). Although, differences can be observed at higher temperatures, the tendency of the temperature values are still similar. Using the temperature dependent conductivity, an acceptable agreement can be achived with the experimental data.

The movement of the evaporative front is shown in the Fig. 6.a-b. The location of the evaporative front is indicated by circles. The characteristic of this moving front shows exponential behaviour. It agrees with our assumptions that the evaporation process recedes towards to the center, which is caused by the absorptive power of the water. The radiative term was neglected in the wet zone, because the moisture content absorbs the main part of the radiation entering in the wood. The influance of the radiation effect has importance in the dry zone, where the moisture content is not continuous in reality.

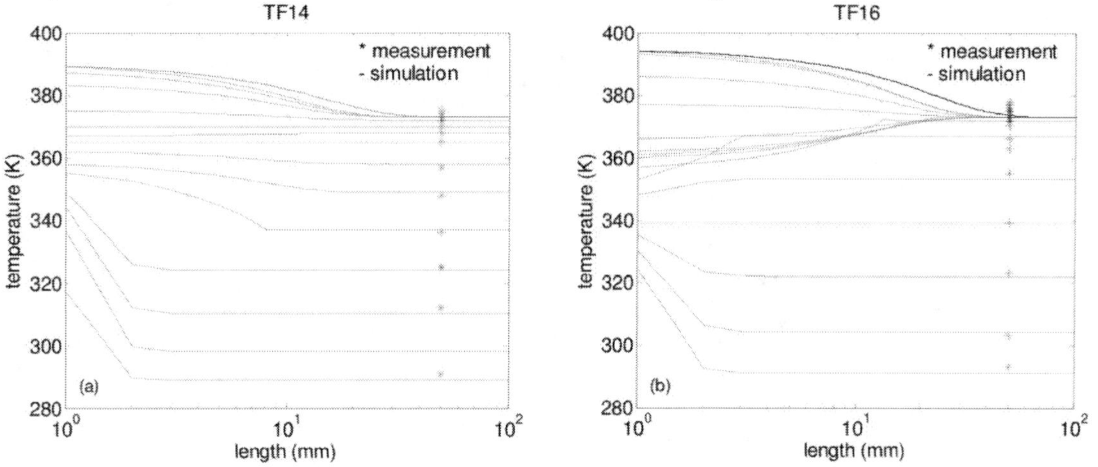

Figure 5. Comparison of the simulated and measured temperature field in the mid-point of the spruce wood in the computational domain

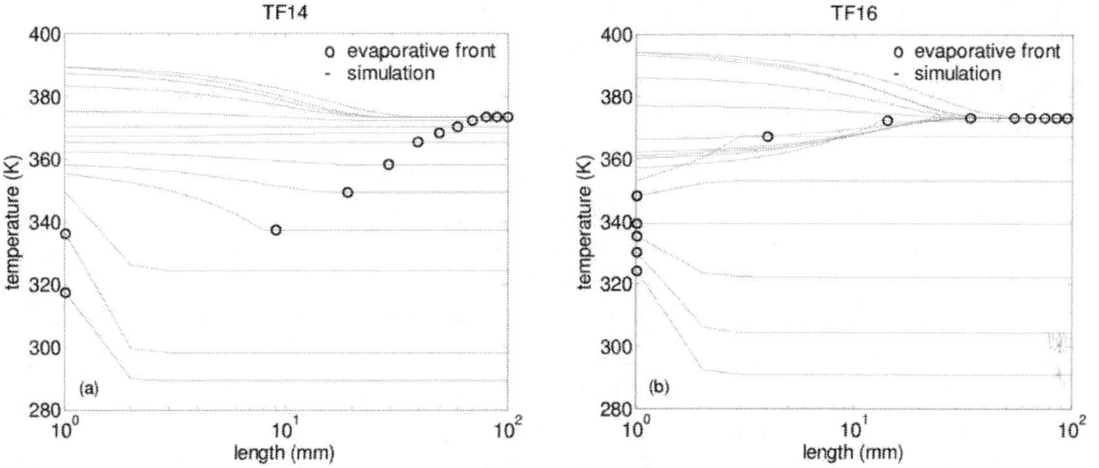

Figure 6. The simulated temperature distribution and the receding evaporative front in the computational domain

Conclusion

The present paper deals with the drying process of spruce wood in a newly developed IR drying furnace according to the patent Nr. PO 700 453. The required frequency field is ensured by heating wires using modificator layers at the optimal heating rate. The IR drying process is developed to have an effective time and energy save technology. A one-dimensional model was set up to investigate numerically the heat transfer in the wood slab. The finite difference method was used to obtain an approximate solution of the temperature profile using explicit and implicit schemes. The numerically predicted results were compared to temperature measurements concerning different process parameters. An acceptably good agreement with experimental data were obtained. For the future work, it is essential to take into account that the optimal heating rate is strongly depending on the inhomogeneous structure of the wood. This research can be a basis of the investigation of further woods.

Acknowledgements

This work is financially supported by the Kotla-Project for which the authors kindly acknowledge. The authors wish to express their appreciation to Sedo Group Ltd., Askada Ltd., and Kentech Ltd. for their industrial contribution in the experimental work as well.

References

1. Rattanedecho, P. (2006), "The Simulation of Microwave Heating of Wood Using a Rectangularwave Guide-Influence of Frequency and Sample Size," *Chemical Engineering Science*, March, pp. 4798-4811.
2. Haghi, A. K. and Ghandzadeh, H. (2006), "Experimental study on combined inrfared and microwave drying of porous media with particular application in wood industry," *Int. J. of Applied Mechanics and Engineering*, Vol. 11, pp. 985-991.
3. Oloyede, A. and Groombridge P. (2000), "The Influence of Microwave Heating on the Mechanical Properties of Wood," *Journal of Materials Processing Technology*, August, pp. 67-73.
4. Hansson L. and Antti A.L. (2002), "The effect of microwave drying on Norway spruce woods strength: a comparison with conventional drying," *Journal of Materials Processing Technology*, November, pp. 41-50.
5. Turner, I. W. and Perré, P. (2004), "Vacuum drying of wood with radiative heating: II. Comparison between theory and experiment," *AIChE Journal*, Vol. 50, pp. 108-118.
6. Bucki, M. and Perré, P. (2003), "Radio requency/vacuum drying of wood: A comprihensive 2-D computational model on the board's scale," *8th International IUFRO Wood Drying Conference*, pp. 33-38.
7. Awadalla, H.S.F. and El-Dib, A.F. and Mohamad, M.A. and Reuss M. and Hussein. H.M.S. (2004), "Mathematical modelling and experimental verification of wood drying process," *Energy Conversion and Management*, pp. 197-207.
8. Shusheng, P. and Keey, R. B. (1995), "Modelling the temperature profiles within board during the high-temperature drying of Pinus radiata timber: the influence of airflow reversals," *Int. J. Heat Mass Transfer*, pp. 189-205.
9. Surasani, V.K. and Metzger, T. and Tsotsas E. (2008), "Consideration of heat transfer in pore network modelling of convective drying," *International Journal of Heat and Mass Transfer*, Vol. 51, pp. 2506-2518.
10. Erriguible, A. and Bernada, P. and Couture, F. and Roques, M.-A. (2007), "Simulation of vacuum drying by coupling models," *Chemical Engineering and Processing*, Vol. 46, pp. 1274-1285.
11. Thunman, H. and Lecker, B. (2002), "Thermal conductivity of wood – models for different stages of combustion," *Biomass and Bioenergy*, Vol. 23, pp. 47-54.
12. Czibere, T. (1998), "Vezetéses hőátvitel (in Hungarian)," Miskolc University Press
13. Forest Products Laboratory (2002), "Wood Handbook – Wood as an engineering material," *USDA Forest Service*, Madison, Wisconsin

665b Particle and Bubble Interactions Visualized In Flotation Systems

Steve R. Duke, Gopal Krishnagopalan, and Zachery Emerson. Chemical Engineering, Auburn University, 212 Ross Hall, Auburn University, AL 36849-5127

Flotation separation processes important to paper recycling were studied using high-speed and high-magnification visualization techniques. Toner ink and model particle interactions with stationary and flowing bubbles in several system chemistries and flow facilities are discussed. Visualizations and image analysis results relate particle and bubble adhesion to particle and bubble properties, surfactants, calcium concentration, and flow parameters. Estimates for the probabilities of particle to bubble collisions, adhesion, and stable attachment determined experimentally were found to differ from probabilities based on predictive modeling.

665c Integrated Biorefinery Approach in a Conventional Kraft Pulp Mill Using Hemicellulose Pre-Extraction for Ethanol Production

Hua-Jiang Huang, **Shri Ramaswamy**, Waleed Wafa Al-Dajani, and Ulrike Tschirner. Department of Bioproducts and Biosystems Engineering, University of Minnesota, 2004 Folwell Ave, Saint Paul, MN 55108

The conversion of biomass to biofuels has gained more and more interest recently because of the limited source of fossil fuels, greenhouse gases contributing to global warming due to the use of fossil fuels, the increase demand of energy in the world, especially in developing countries, and the need for renewable energy.

In the near future, integrated biorefineries (IBR) based on pulp mills could become the most important lignocellulosic biorefineries due to availability of additional forest residues, synergies with existing infrastructure and potential capital and operating cost benefits. Therefore, it is important to analyze and compare the process techno-economic performance of conventional Kraft pulping and pulp mill-based biorefinery with pre-extraction of hemicellulose for ethanol production. Experimental data has been used in developing the modified model with hemicellulose pre-extraction.

The objective of this presentation is to perform simulation of a conventional Kraft pulping process using the simulator WINGEMS and a modified pulp mill, i.e., an integrated biorefinery based on pulp mill with pre-extraction of hemicellulose for ethanol production by using Aspen Plus, and then comparison of these two processes by techno-economic analysis, in terms of energy (heat and power) consumption or generation, total capital costs, operating costs, and production cost, etc.

The comparative study shows that the integrated biorefinery based on pulp mill using hemicellulose pre-extraction could bring added value and much needed increased revenue for pulp and paper industry.

665d Degradation of Polylactide Films in Water and Moisture

Yuyang Liu, Ulrike Tschirner, and Shri Ramaswamy. Department of Bioproducts and Biosystems Engineering, University of Minnesota, 2004 Folwell Avenue, St. Paul, MN 55108

Poly Lactic Acid (or polylactide, PLA) is a biodegradable polymer with a wide range of commodity applications, such as medical, textile and packaging materials. The degradation characteristics of PLA polymers in water and moisture are extremely important during manufacturing, shipping, and storage and end-use applications of PLA materials. The aim of this research is to better understand how the degradation depends on physical structures of PLA polymers and the environmental parameters.

PLA polymer films with different thickness (2mm, 50μ m, and 25 μ m) and crystallinity (amorphous and semi-crystalline) were degraded in water and moisture (RH) environments to understand the role of physical structures and environmental parameters in the degradation process of lactic acid based polymers. The degradation studies were conducted in water and moisture environment maintained at 60 and 80°C. Various analytical techniques were applied to monitor the changes in the rate/or mechanism of degradation. Weight loss, pH value, titration, and molecular weights and molecular weight distributions were measured to evaluate the extent of degradation. Size exclusion high performance liquid chromatography (SEC-HPLC) was used as a quantitative tool to analyze the products from the degradation. The results showed that the rate of degradation high at the beginning (first few days) decrease significantly over time. Also, the results suggest a dominant contribution by the end groups and hydrolysis of ester linkage as the primary mechanism of degradation of PLA. Spatial variations and gradient in rate of degradation of PLA matrix was also investigated. On the basis of these data, a probable route of PLA degradation has been suggested.

AUTHOR INDEX